Springer Series in Statistics

Andrews/Herzberg: Data: A Collection of Problems from Many Fields for the Student and Research Worker.
Anscombe: Computing in Statistical Science through APL.
Berger: Statistical Decision Theory and Bayesian Analysis, 2nd edition.
Brémaud: Point Processes and Queues: Martingale Dynamics.
Brockwell/Davis: Time Series: Theory and Methods.
Daley/Vere-Jones: An Introduction to the Theory of Point Processes.
Dzhaparidze: Parameter Estimation and Hypothesis Testing in Spectral Analysis of Stationary Time Series.
Farrell: Multivariate Calculation.
Fienberg/Hoaglin/Kruskal/Tanur (eds.): A Statistical Model: Frederick Mosteller's Contributions to Statistics, Science, and Public Policy.
Goodman/Kruskal: Measures of Association for Cross Classifications.
Hartigan: Bayes Theory.
Heyer: Theory of Statistical Experiments.
Jolliffe: Principal Component Analysis.
Kres: Statistical Tables for Multivariate Analysis.
Leadbetter/Lindgren/Rootzén: Extremes and Related Properties of Random Sequences and Processes.
Le Cam: Asymptotic Methods in Statistical Decision Theory.
Manoukian: Modern Concepts and Theorems of Mathematical Statistics.
Miller, Jr.: Simultaneous Statistical Inference, 2nd edition.
Mosteller/Wallace: Applied Bayesian and Classical Inference: The Case of *The Federalist Papers*.
Pollard: Convergence of Stochastic Processes.
Pratt/Gibbons: Concepts of Nonparametric Theory.
Read/Cressie: Goodness-of-Fit Statistics for Discrete Multivariate Data.
Reiss: Approximate Distributions of Order Statistics: With Applications to Nonparametric Statistics.
Ross: Nonlinear Estimation.
Sachs: Applied Statistics. A Handbook of Techniques, 2nd edition.
Seneta: Non-Negative Matrices and Markov Chains.
*Siegmund:*Sequential Analysis: Tests and Confidence Intervals.
Tong: The Multivariate Normal Distribution.
Vapnik: Estimation of Dependences Based on Empirical Data.
West/Harrison: Bayesian Forecasting and Dynamic Models.
Wolter: Introduction to Variance Estimation.
Yaglom: Correlation Theory of Stationary and Related Random Functions I: Basic Results.
Yaglom: Correlation Theory of Stationary and Related Random Functions II: Supplementary Notes and References.

Springer Series in Statistics
Perspectives in Statistics

Advisors
J. Berger, S. Fienberg, J. Gani,
K. Krickeberg, I. Olkin, B. Singer

Edited by
S.E. Fienberg D.C. Hoaglin
W.H. Kruskal J.M. Tanur

With the Collaboration of
Cleo Youtz

A Statistical Model

Frederick Mosteller's Contributions to Statistics, Science, and Public Policy

Springer-Verlag
New York Berlin Heidelberg
London Paris Tokyo Hong Kong

Preface

When planning for this volume began early in 1984, our intent was to complete everything in time for Fred Mosteller's 70th birthday on December 24, 1986. The alert reader will note that we are off by only about three years—Fred celebrates his 73rd birthday today. It would be easy to try to blame this delay on others, including Fred himself. After all, his research productivity has continued at such a substantial rate that we found ourselves frequently updating the bibliography and several chapters. (Fred remarked to one of us earlier this summer: "I had one of my most productive years in 1988!") The real delays lay with us, and they came, at least in part, because we had set out to prepare something other than just another festschrift volume, and thus we were demanding of our collaborators and contributors. As someone remarked at Fred's 70th birthday party: "What you folks needed was Fred Mosteller as the editor if you wanted to get the job done properly and on time!" Anyone who has worked with Fred over the years will understand the truth of this comment.

Fred has an extremely long list of collaborators and coauthors, and many of them contributed material for this volume. Their diverse perspectives have combined to give a rich picture of Fred's many activities and contributions. After a brief biography and a detailed bibliography, successive chapters discuss his contributions to science, to mathematical statistics, to methodology and applications, and as an educator. A further chapter focuses on activities that have had some connection with Harvard University, some of which are discussed in other chapters and some of which are not. Many of the projects that Fred has led or participated in have produced books. The final chapter assembles 21 new book reviews that examine these efforts and their impact.

One of the contributions that we gathered never quite fit into our outline, but it is important, and we include part of it here. Virginia, Fred's wife, shared with us recollections of his graduate student days:

> During the early days before the War Fred seemed uncertain of his ability to finish the work toward a degree. He seemed to feel that he was not well prepared in various basic math courses.

> He was pleased when he was asked to do some war research. He enjoyed the various problems posed to them for research, and often it was necessary to do a quick study and provide partial answers as best they could, often involving late hours, which did not bother him at all. What I did not realize at the time was that working with others in his field on this basis was giving him a kind of training that stood him in good stead when the War was over and he could finally sit down to write his thesis. He seemed to have confidence that he would do it eventually, and it made our lives a lot more serene....
>
> As people came back to Princeton after the War, almost everyone had some adjustments to make in their lives—new jobs to do, new places to live, etc. I was amazed at how many people came to Fred for advice. He passed it off as a result of his being one of the oldest graduate students around; but it continued long after that, and I realized that people seemed to recognize that he had a kind of common sense that could help solve some of the problems in their lives. I think he is sought out today for the same kind of advice.

We owe special thanks to two individuals who have labored with us on this volume from the beginning. Cleo Youtz has worked tirelessly on organizing Fred's bibliography and in helping us to get many details in various chapters correct. Cleo has worked with Fred since the 1950s and often has been the only one who knew who had worked on projects with Fred and how we could get in touch with them. She has also contributed text in several places, helped to round up old photographs, and, but for her modesty, would have been listed as a coeditor. Margaret L. Smykla prepared innumerable drafts of various chapters, as well as the final camera-ready manuscript, on the computing network of the Department of Statistics at Carnegie Mellon University.

Both of us have worked with Fred on several book projects over the years, and one of the many lessons we have learned from him is how to date the preface of a book. Unfortunately for Fred, the hard work of choosing a date for the preface has been undone on several of his books by overzealous copy-editors, who have deleted the day of the month (as in the prefaces for the two editions of *The Federalist* book with David Wallace). We have had no such problem here.

We end these prefatory remarks by joining in a 70th birthday toast prepared by yet another of Fred's colleagues, William T. Golden, long-time Treasurer of the American Association for the Advancement of Science, who worked with Fred during his service as President of that organization:

"As you achieve three score and ten,
I reaffirm, dear wise man Fred,
Respect, devotion, faith; and then
Add hope for years and years," he said.

With that fine toast we drink our fill
And send three cheers from your friends.

December 24, 1989

Stephen E. Fienberg
David C. Hoaglin

Contents

List of Contributors

JOHN C. BAILAR III
McGill University

DONALD M. BERWICK
Harvard Community Health Plan

YVONNE M. BISHOP
U.S. Department of Energy

JOHN P. BUNKER
Stanford University School of Medicine

WILLIAM D. CAREY
Executive Officer, AAAS (Ret.)

RALPH B. D'AGOSTINO
Boston University

JAMES A. DAVIS
Harvard University

PERSI DIACONIS
Harvard University

JANET D. ELASHOFF
University of California, Los Angeles

JOHN D. EMERSON
Middlebury College

DORIS R. ENTWISLE
Johns Hopkins University

WILLIAM B. FAIRLEY
Analysis and Inference, Inc.

STEPHEN E. FIENBERG
Carnegie Mellon University

HARVEY V. FINEBERG
Harvard School of Public Health

WILLIAM T. GOLDEN
40 Wall Street, New York, NY

HOWARD H. HIATT
Harvard Medical School

DAVID C. HOAGLIN
Harvard University

PAUL W. HOLLAND
Educational Testing Service

RAY HYMAN
University of Oregon

GUDMUND R. IVERSEN
Swarthmore College

LYLE V. JONES
University of North Carolina

JOEL C. KLEINMAN
National Center for Health Statistics

WILLIAM KRUSKAL
University of Chicago

STEPHEN W. LAGAKOS
Harvard School of Public Health

NAN LAIRD
Harvard School of Public Health

LOUIS LASAGNA
Tufts University

ERICH LEHMANN
University of California, Berkeley

THOMAS A. LEHRER
University of California, Santa Cruz

RICHARD J. LIGHT
Harvard University

RICHARD F. LINK
Richard F. Link & Associates, Inc.

THOMAS A. LOUIS
University of Minnesota School of Public Health

LINCOLN E. MOSES
Stanford University

GALE MOSTELLER
Federal Trade Commission

VIRGINIA MOSTELLER
Belmont, MA

WILLIAM MOSTELLER
Systems Center, Inc.

JOSEPH I. NAUS
Rutgers University

EMANUEL PARZEN
Texas A&M University

JOHN W. PRATT
Harvard University

T.E. RAGHUNATHAN
University of Washington

JOHN S. REED
Russell Sage Foundation

ROBERT ROSENTHAL
Harvard University

DONALD B. RUBIN
Harvard University

MICHAEL A. STOTO
Institute of Medicine, Washington, D.C.

MICHAEL SUTHERLAND
University of Massachusetts, Amherst

JUDITH M.TANUR
State University of New York, Stony Brook

STEPHEN R. THOMAS
Commonwealth Fund

JOHN W. TUKEY
Princeton University

LEROY D. VANDAM
Harvard Medical School

JAMES H. WARE
Harvard School of Public Health

W. ALLEN WALLIS
U.S. Department of State

SANFORD WEISBERG
University of Minnesota

CLEO YOUTZ
Harvard University

MARVIN ZELEN
Harvard School of Public Health

Chapter 1

BIOGRAPHY

John W. Tukey

Frederick Mosteller was born Charles Frederick Mosteller in Clarksburg, West Virginia, the son of Helen Kelley Mosteller and William Roy Mosteller. His family soon moved to Parkersburg, W. Va. and then to Wilkinsburg, a suburb of Pittsburgh, Pennsylvania. They later moved to Pittsburgh and then back to Wilkinsburg. He went to Schenley High School in Pittsburgh—where he and a friend started a chess club—and to Carnegie Institute of Technology (now Carnegie Mellon University)—where he was on the chess team. He worked summers for his father, a highway builder, to earn money for college. When it rained, the work force laid off and played poker. To preserve his college funds, Fred had to learn rapidly to play poker well.

When Fred was a sophomore, he rode the street car to college daily; Virginia Gilroy, a freshman, rode the same street car—more below. He survived a very practical engineering education—there is a rumor about courses in carpentry, welding, masonry, and bricklaying—but his main interest was in mathematics. A "how many ways" problem in a physical measurements class (recounted in Collins, 1986) led him into the influence of Edwin G. Olds, where he became committed to Statistics and took an Sc.M. degree in 1939. Fred has always asserted that a great teacher is one who trains someone even better than himself. No one can doubt that Ed Olds, who also developed David Wallace and many others, met this standard.

I first saw Fred in the fall of 1939, when he came to Princeton's Graduate College as a graduate student in mathematics, where I had just gotten a Ph.D. in topology. Being a statistician among pure mathematicians was not easy, but Fred survived. Fred's work with Sam Wilks, helping him as Editor of the *Annals of Mathematical Statistics*, played a significant role in his statistical education. Thus it is specially fitting that he received the ASA Samuel S. Wilks award in 1986.

In the spring of 1941, Virginia Gilroy came to Princeton to visit Fred. Will (W.J.) Dixon persuaded her to stay over and talk to Merrill M. Flood, who needed secretarial staff. So Fred and Virginia were married in May 1941, and Virginia came to work at Fire Control Research, where I was already part-time. Fred worked hard, first in part-time war work combined with teaching, and then in full-time war work, ending up in SRG–Pjr—the Morningside Heights (Upper Manhattan) branch of Sam Wilks's Statistical phasis on careful exposition and from his willingness to work in unclear fields. As a collaborator, I notice that he wants to have things look more mathematical. I am unclear about whether this is an unmelted residue of taking mathematicians as a reference group or just being a very shrewd judge of the marketplace.

Selected honors deserve mention here, including four honorary degrees (Chicago, 1973; Carnegie Mellon, 1974; Yale, 1981; Wesleyan, 1983), a Guggenheim Fellowship (1969–1970), election as an honorary fellow of the Royal Statistical Society, years at the Center for Advanced Study in the Behavioral Sciences (1962–1963) and as a Miller Research Professor at Berkeley (1974–1975), and appearance as the Hitchcock Professor (The University of California's outstanding lectureship) in 1984–1985.

I had the pleasure of serving with him on the Analysis Advisory Committee of the National Assessment of Educational Progress and on the President's Commission on Federal Statistics, of which he was Vice-Chairman. In these two activities, as in so many others, he was always a quiet, thoughtful contributor. His input was important at the time; but, even more important in retrospect, his influence was always wise and remarkably tolerant, often by recognizing good things that others might have missed.

He has been elected to each of the "big three" honorary academies: the National Academy of Sciences, the American Philosophical Society held at Philadelphia for the promotion of useful knowledge, and the American Academy of Arts and Sciences. He has served well and effectively on the NAS-related Institute of Medicine. He has received prizes from the Chicago Chapter of ASA, the Evaluation Research Society, and the Council for Applied Social Research; he has been president of the Psychometric Society (1957–1958), the American Statistical Association (1967), the Institute of Mathematical Statistics (1974–1975), and the American Association for the Advancement of Science (1980), as well as being chairman of the board of directors of the Social Science Research Council (1965–1969) and vice-president of the International Statistical Institute (1985–1987). He served long and well on the board of directors of the Russell Sage Foundation (1964–1985).

When we look back at how early some of Fred's interests began—unpublished work (with Frederick Williams) on the *Federalist* papers and on polling in the early '40s, medical collaboration by 1950—and how long many of these interests have continued, who can believe that formal retirement (at least from administration and teaching) could cut off either his manifold activities or his interactions with many collaborators and students. It is said of the great anthropologist, Alfred Kroeber, that what he did after he retired would have filled an ordinary man's career. How can his friends and admirers expect less of Fred?

Reference

Collins, C. (1986). Frederick Mosteller: Statistician at Large. *Carnegie Mellon Magazine*, Summer 1986, pp. 9–10.

Chapter 2

BIBLIOGRAPHY

This annotated bibliography of Frederick Mosteller's writings has four sections: books, numbered B1 through B56; papers, numbered P1 through P182; miscellaneous writings, numbered M1 through M36; and reviews, numbered R1 through R25. Ordinarily we list publications chronologically within each of these sections. We deviate from that ordering to bring together closely related books. For example, B19 (1984) is an expanded edition of B18 (1964).

We list Fred's name as it appeared on the publication, usually "Frederick Mosteller." For works with other authors, we list the names in the order that they appeared on the publication. For publications by a committee or panel, we list the members.

If, in addition to being one of the editors of a book, Fred is author or coauthor of a chapter, we list that chapter under the book but not among the papers. For example, B47 was edited by Hoaglin, Mosteller, and Tukey, and they also coauthored Chapter 9. We therefore list Chapter 9 under B47.

We give any additional information that we have, such as new editions, different printings, and translations. We realize that our information about translations is not complete, nor do we always know of reprinting of papers or chapters. When a paper is an abridgment of a chapter or another paper, we indicate that.

BOOKS

B1 Hadley Cantril and Research Associates in the Office of Public Opinion Research, Princeton University. *Gauging Public Opinion.* Princeton: Princeton University Press, 1944. Second printing, 1947.

- Chapter IV. Frederick Mosteller and Hadley Cantril. "The use and value of a battery of questions." pp. 66–73.
- Chapter VII. Frederick Mosteller. "The reliability of interviewers' ratings." pp. 98–106.
- Chapter VIII. William Salstrom with Daniel Katz, Donald Rugg, Frederick Mosteller, and Frederick Williams. "Interviewer bias and rapport." pp. 107–118.
- Chapter XIV. Frederick Williams and Frederick Mosteller. "Education and economic status as determinants of opinion." pp. 195–208.

- Appendix II. Frederick Mosteller. "Correcting for interviewer bias." pp. 286–288.

- Appendix III. Frederick Mosteller. "Sampling and breakdowns: technical notes." pp. 288–296.

- Appendix IV. Frederick Mosteller. "Charts indicating confidence limits and critical differences between percentages." pp. 297–301.

B2 H.A. Freeman, Milton Friedman, Frederick Mosteller, and W. Allen Wallis, editors. *Sampling Inspection: Principles, Procedures, and Tables for Single, Double, and Sequential Sampling in Acceptance Inspection and Quality Control Based on Percent Defective.* New York: McGraw-Hill, 1948.

By the Statistical Research Group, Columbia University, Applied Mathematics Panel, Office of Scientific Research and Development.

"The manuscript was prepared by H.A. Freeman, Milton Friedman, Frederick Mosteller, Leonard J. Savage, David H. Schwartz, and W. Allen Wallis. ...Chaps. 6 and 13 are primarily the work of Messrs. Mosteller and Schwartz ..." (p. vii).

- Chapter 6. Frederick Mosteller and David H. Schwartz. "Use of sampling inspection for quality control." pp. 55–68.

- Chapter 13. Frederick Mosteller and David H. Schwartz. "Application of the standard procedure to control sampling." pp. 135–136.

B3 Frederick Mosteller, Herbert Hyman, Philip J. McCarthy, Eli S. Marks, and David B. Truman, with the collaboration of Leonard W. Doob, Duncan MacRae, Jr., Frederick F. Stephan, Samuel A. Stouffer, and S.S. Wilks. *The Pre-election Polls of 1948: Report to the Committee on Analysis of Pre-election Polls and Forecasts.* New York: Social Science Research Council, Bulletin 60, 1949.

Members of Committee on Analysis of Pre-election Polls and Forecasts: S.S. Wilks, Chairman, Frederick F. Stephan, Executive Secretary, James Phinney Baxter, 3rd, Philip M. Hauser, Carl I. Hovland, V.O. Key, Isador Lubin, Frank Stanton, and Samuel A. Stouffer.

Although this report was a group effort, preparation of each chapter was assigned to a particular person or persons.

- Chapter V. Prepared by Frederick Mosteller. "Measuring the error." pp. 54–80.

B4 William G. Cochran, Frederick Mosteller, and John W. Tukey, with the assistance of W.O. Jenkins. *Statistical Problems of the Kinsey Report on Sexual Behavior in the Human Male: A Report of the American Statistical Association Committee to Advise the National Research Council Committee for Research in Problems of Sex.* Washington, D.C.: The American Statistical Association, 1954.

Two parts of this book were published earlier—papers P33 and P34.

B5 Robert R. Bush and Frederick Mosteller. *Stochastic Models for Learning.* New York: Wiley, 1955.

B6 A group of the Commission on Mathematics of the College Entrance Examination Board. *Introductory Probability and Statistical Inference for Secondary Schools: An Experimental Course.* Preliminary edition. New York: College Entrance Examination Board, 1957.

Members of the group: Edwin C. Douglas, Frederick Mosteller, Richard S. Pieters, Donald E. Richmond, Robert E.K. Rourke, George B. Thomas, Jr., and Samuel S. Wilks. These names were inadvertently omitted from this edition.

B7 Revised edition of B6, 1959.

B8 A group of the Commission on Mathematics of the College Entrance Examination Board. *Teachers' Notes and Answer Guide: Supplementary material for the revised preliminary edition of "Introductory Probability and Statistical Inference."* New York: College Entrance Examination Board, 1959.

Members of the group: Frederick Mosteller, Richard S. Pieters, Robert E.K. Rourke, George B. Thomas, Jr., and Samuel S. Wilks.

B7 and B8 were translated by Prof. Marta C. Valincq into Spanish, and published by Comision de Educacion Estadistica del Instituto Interamericano de Estadistica Rosario (Rep. Argentina), 1961.

B9 Commission on Mathematics. *Program for College Preparatory Mathematics.* Report of the Commission. New York: College Entrance Examination Board, 1959.

Members of the Commission on Mathematics: Albert W. Tucker, Chairman, Carl B. Allendoerfer, Edwin C. Douglas, Howard F. Fehr, Martha Hildebrandt, Albert E. Meder, Jr., Morris Meister, Frederick Mosteller, Eugene P. Northrop, Ernest R. Ranucci, Robert E.K. Rourke, George B. Thomas, Jr., Henry Van Engen, and Samuel S. Wilks.

B16 Frederick Mosteller, Robert E.K. Rourke, and George B. Thomas, Jr. Reading, MA: Addison-Wesley.

B10 *Probability with Statistical Applications*, 1961. Second printing, 1965.

B11 *Probability with Statistical Applications*, Second Edition, 1970.

Extensive revision of B10.

World Student Series Edition, Third Printing, 1973, of B11. Published for the Open University in England.

B12 *Probability and Statistics*, Official textbook for Continental Classroom, 1961.

Derived from B10 by abridgment and slight rewriting. Also translated into Turkish.

B13 *Probability: A First Course*, 1961.

Derived from B10 by abridgment and slight rewriting. Also translated into Russian. And a Bulgarian translation into Russian. Sofia, Bulgaria: J. Stoyanov, 1975.

B14 *Probability: A First Course*, Second Edition, 1970.

B15 *Teacher's Manual for Probability: A First Course*, 1961.

B16 *Instructor's Manual to Accompany Probability with Statistical Applications, Second Edition, and Probability: A First Course, Second Edition*, 1970.

B17 Frederick Mosteller, Keewhan Choi, and Joseph Sedransk. *A Catalogue Survey of College Mathematics Courses.* Mathematical Association of America, Committee on the Undergraduate Program in Mathematics, Report Number 4, December 1961.

B18 Frederick Mosteller and David L. Wallace. *Inference and Disputed Authorship: The Federalist.* Reading, MA: Addison-Wesley, 1964.

Related papers: P53, P58, and P159.

B19 Frederick Mosteller and David L. Wallace. *Applied Bayesian and Classical Inference: The Case of The Federalist Papers.* New York: Springer-Verlag, 1984.

Second Edition of B18. A new chapter dealing with authorship work published from about 1969 to 1983 was added in the second edition. A new, lengthy Analytic Table of Contents replaced the original Table of Contents.

B20 Panel on Mathematics for the Biological, Management, and Social Sciences. Committee on the Undergraduate Program in Mathematics, Mathematical Association of America. *Tentative Recommendations for the Undergraduate Mathematics Program of Students in the Biological, Management and Social Sciences.* Berkeley, CA: CUPM Central Office, 1964.

Panel members: John G. Kemeny, Chairman, Joseph Berger, Robert R. Bush, David Gale, Samuel Goldberg, Harold Kuhn, Frederick Mosteller, Theodor D. Sterling, Gerald L. Thompson, Robert M. Thrall, A.W. Tucker, and Geoffrey S. Watson.

B21 Frederick Mosteller. *Fifty Challenging Problems in Probability with Solutions.* Reading, MA: Addison-Wesley, 1965.

Also translated into Russian, 1975.

A second edition of the Russian translation was published in 1985. For this edition, Frederick Mosteller provided a new problem, "Distribution of prime divisors."

B22 Frederick Mosteller. *Fifty Challenging Problems in Probability with Solutions.* New York: Dover, 1987.

Reissue of B21.

B23 Panel on Undergraduate Education in Mathematics of the Committee on Support of Research in the Mathematical Sciences of the National Research Council. *The Mathematical Sciences: Undergraduate Education.* Washington, D.C.: National Academy of Sciences, 1968.

Panel members: John G. Kemeny, Grace E. Bates, D.E. Christie, Llayron Clarkson, George Handelman, Frederick Mosteller, Henry Pollak, Hartley Rogers, John Toll, Robert Wisner, and Truman A. Botts.

B24 John P. Bunker, William H. Forrest, Jr., Frederick Mosteller, and Leroy D. Vandam, editors. *The National Halothane Study: A Study of the Possible Association between Halothane Anesthesia and Postoperative Hepatic Necrosis.* Report of The Subcommittee on the National Halothane Study, of the Committee on Anesthesia, Division of Medical Sciences, National Academy of Sciences-National Research Council. National Institutes of Health, National Institute of General Medical Sciences. Washington, D.C.: U.S. Government Printing Office, 1969.

Related papers: P66 and P72.

- Part IV, Chapter 1. Byron W. Brown, Jr., Frederick Mosteller, Lincoln E. Moses, and W. Morven Gentleman. "Introduction to the study of death rates." pp. 183–187.

- Part IV, Appendix 3 to Chapter 2. Frederick Mosteller. "Estimation of death rates." pp. 234–235.

- Part IV, Chapter 3. Yvonne M.M. Bishop and Frederick Mosteller. "Smoothed contingency-table analysis." pp. 237–272.

- Part IV, Chapter 8. Lincoln E. Moses and Frederick Mosteller. "Afterword for the study of death rates." pp. 395–408.

B25 The President's Commission on Federal Statistics. *Federal Statistics: Report of the President's Commission, Vol. I.* Washington, D.C.: U.S. Government Printing Office, 1971.

Members of the Commission: W. Allen Wallis, Chairman, Frederick Mosteller, Vice-Chairman, Ansley J. Coale, Paul M. Densen, Solomon Fabricant, W. Braddock Hickman, William Kruskal, Robert D. Fisher, Stanley Lebergott, Richard M. Scammon, William H. Shaw, James A. Suffridge, John W. Tukey, and Frank D. Stella.

The Staff: Daniel B. Rathbun, Executive Director, Paul Feldman, Deputy Executive Director, and Norman V. Breckner, Assistant Director.

B26 The President's Commission on Federal Statistics. *Federal Statistics: Report of the President's Commission, Vol. II.* Washington, D.C.: U.S. Government Printing Office, 1971.

- Chapter 6. Richard J. Light, Frederick Mosteller, and Herbert S. Winokur, Jr. "Using controlled field studies to improve public policy." pp. 367–398.

B27–B31 Judith M. Tanur and members: Frederick Mosteller, Chairman, William H. Kruskal, Richard F. Link, Richard S. Pieters, and Gerald R. Rising of the Joint Committee on the Curriculum in Statistics and Probability of the American Statistical Association and the National Council of Teachers of Mathematics, editors.

These comprise essays by many authors, and the same essay may appear in more than one of the books.

B27 *Statistics: A Guide to the Unknown.* San Francisco: Holden-Day, 1972.

B28 Second edition of B27. San Francisco: Holden-Day, 1978. With Erich L. Lehmann, Special Editor. Re-issued: Monterey, CA: Wadsworth & Brooks/Cole, 1985.

B28a Third edition of B27. Judith M. Tanur, Frederick Mosteller, William H. Kruskal, Erich L. Lehmann, Richard F. Link, Richard S. Pieters, and Gerald R. Rising, editors. Pacific Grove, CA: Wadsworth & Brooks/Cole, 1989.

B29 *Statistics: A Guide to Business and Economics.* San Francisco: Holden-Day, 1976. With E.L. Lehmann, Special Editor.

B30 *Statistics: A Guide to the Biological and Health Sciences.* San Francisco: Holden-Day, 1977. With E.L. Lehmann, Special Editor.

B31 *Statistics: A Guide to Political and Social Issues.* San Francisco: Holden-Day, 1977. With E.L. Lehmann, Special Editor.

- Frederick Mosteller. "Foreword." In B27, pp. viii–x; B28, pp. ix–xi; B28a, pp. ix–x; B29, pp. viii–x.

- Lincoln E. Moses and Frederick Mosteller. "Safety of anesthetics." In B27, pp. 14–22; B28, pp. 16–25; B28a, pp. 15–24; B30, pp. 101–110.

- Frederick Mosteller and David L. Wallace. "Deciding authorship." In B27, pp. 164–175; B28, pp. 207–219; B28a, pp. 115–125; B31, pp. 78–90.

- John P. Gilbert, Bucknam McPeek, and Frederick Mosteller. "How frequently do innovations succeed in surgery and anesthesia?" In B28, pp. 45–58; B30, pp. 51–64.

- John P. Gilbert, Richard J. Light, and Frederick Mosteller. "How well do social innovations work?" In B28, pp. 125–138; B31, pp. 47–60.

B32 Frederick Mosteller and Daniel P. Moynihan, editors. *On Equality of Educational Opportunity*: Papers deriving from the Harvard University Faculty Seminar on the Coleman Report. New York: Random House, 1972.

- Chapter 1. Frederick Mosteller and Daniel P. Moynihan. "A pathbreaking report." pp. 3–66.

- Chapter 8. John P. Gilbert and Frederick Mosteller. "The urgent need for experimentation." pp. 371–383.

B33a Frederick Mosteller and Robert E.K. Rourke. *Sturdy Statistics: Nonparametrics and Order Statistics.* Reading, MA: Addison-Wesley, 1973.

B33b Frederick Mosteller and Robert E.K. Rourke. *Solutions Manual for Sturdy Statistics: Nonparametrics and Order Statistics.* Reading, MA: Addison-Wesley, 1973.

B34–B37 Frederick Mosteller, William H. Kruskal, Richard F. Link, Richard S. Pieters, and Gerald R. Rising, editors. The Joint Committee on the Curriculum in Statistics and Probability of the American Statistical Association and the National Council of Teachers of Mathematics. Reading, MA: Addison-Wesley, 1973.

Parts of these books were translated into Japanese, 1979.

B34 *Statistics by Example: Exploring Data.* (with the assistance of Martha Zelinka)

- Set 2. Frederick Mosteller. "Fractions on closing stock market prices." pp. 9–13.

B43 Committee for a Planning Study for an Ongoing Study of Costs of Environment-Related Health Effects, Institute of Medicine. *Costs of Environment-Related Health Effects: A Plan for Continuing Study.* Washington, D.C.: National Academy Press, 1981.

Members of the Committee: Kenneth J. Arrow, Chairman, Theodore Cooper, Ralph C. d'Arge, Philip J. Landrigan, Alexander Leaf, Joshua Lederberg, Paul A. Marks, Frederick Mosteller, Evelyn F. Murphy, Robert F. Murray, Don K. Price, Frederick C. Robbins, Anne A. Scitovsky, Irving J. Selikoff, Herman A. Tyroler, Arthur C. Upton, and Richard Zeckhauser.

B44 David C. Hoaglin, Richard J. Light, Bucknam McPeek, Frederick Mosteller, and Michael A. Stoto. *Data for Decisions: Information Strategies for Policymakers.* Cambridge, MA: Abt Books, 1982.

Paperback: Lanham, MD: University Press of America, 1984.

B45–B46 The National Science Board Commission on Precollege Education in Mathematics, Science and Technology. *Educating Americans for the 21st Century: A plan of action for improving mathematics, science and technology education for all American elementary and secondary students so that their achievement is the best in the world by 1995.* Washington, D.C.: National Science Foundation, 1983.

Members of the Commission: William T. Coleman, Jr., Co-Chair, Cecily Cannan Selby, Co-Chair, Lew Allen, Jr., Victoria Bergin, George Burnet, Jr., William H. Cosby, Jr., Daniel J. Evans, Patricia Albjerg Graham, Robert E. Larson, Gerald D. Laubach, Katherine P. Layton, Ruth B. Love, Arturo Madrid II, Frederick Mosteller, M. Joan Parent, Robert W. Parry, Benjamin F. Payton, Joseph E. Rowe, Herbert A. Simon, and John B. Slaughter.

B45 *A Report to the American People and the National Science Board.*

B46 *Source Materials.*

B47 David C. Hoaglin, Frederick Mosteller, and John W. Tukey, editors. *Understanding Robust and Exploratory Data Analysis.* New York: Wiley, 1983.

- Chapter 9. David C. Hoaglin, Frederick Mosteller, and John W. Tukey. "Introduction to more refined estimators." pp. 283–296.

B48 Frederick Mosteller, Stephen E. Fienberg, and Robert E.K. Rourke. *Beginning Statistics with Data Analysis.* Reading, MA: Addison-Wesley, 1983.

Diane L. Griffin and Gale Mosteller. *Solutions Manual to Accompany Mosteller, Fienberg, and Rourke's Beginning Statistics with Data Analysis.* Reading, MA: Addison-Wesley, 1983.

B49 Joseph A. Ingelfinger, Frederick Mosteller, Lawrence A. Thibodeau, and James H. Ware. *Biostatistics in Clinical Medicine.* New York: Macmillan, 1983.

Italian edition: *Biostatistica in Medicina*, translated by Ettore Marubini. Milano: Raffacello Cortina Editore, 1986.

B50 Second edition of B49, 1987.

The second edition has two new chapters—one on life tables, the other on multiple regression.

B51 Lincoln E. Moses and Frederick Mosteller, editors. *Planning and Analysis of Observational Studies*, by William G. Cochran. New York: Wiley, 1983.

At the time of his death, William G. Cochran left an almost completed manuscript on observational studies. Lincoln E. Moses and Frederick Mosteller edited and organized the manuscript, and *Planning and Analysis of Observational Studies* is the result.

B52–B53 Oversight Committee on Radioepidemiologic Tables, Board on Radiation Effects Research, Commission on Life Sciences, National Research Council. Washington, D.C.: National Academy Press, 1984.

Members of the Committee: Frederick Mosteller, Chairman. Jacob I. Fabrikant, R.J. Michael Fry, Stephen W. Lagakos, Anthony B. Miller, Eugene L. Saenger, David Schottenfeld, Elizabeth L. Scott, John R. Van Ryzin, and Edward W. Webster; Stephen L. Brown, Staff Officer, Norman Grossblatt, Editor.

B52 *Assigned Share for Radiation as a Cause of Cancer: Review of Assumptions and Methods for Radioepidemiologic Tables. Interim Report.*

B53 *Assigned Share for Radiation as a Cause of Cancer: Review of Radioepidemiologic Tables Assigning Probabilities of Causation. Final Report.*

Related paper: P157.

B54 David C. Hoaglin, Frederick Mosteller, and John W. Tukey, editors. *Exploring Data Tables, Trends, and Shapes.* New York: Wiley, 1985.

- Chapter 5. Frederick Mosteller and Anita Parunak. "Identifying extreme cells in a sizable contingency table: probabilistic and exploratory approaches." pp. 189–224.

- Chapter 6. Frederick Mosteller, Andrew F. Siegel, Edward Trapido, and Cleo Youtz. "Fitting straight lines by eye." pp. 225–239.

This is an expanded version of paper P133.

B55 Committee for Evaluating Medical Technologies in Clinical Use, Division of Health Promotion and Disease Prevention, Institute of Medicine. *Assessing Medical Technologies.* Washington, D.C.: National Academy Press, 1985.

Members of the Committee: Frederick Mosteller, Chairman, H. David Banta, Stuart Bondurant, Morris F. Collen, Joanne E. Finley, Barbara J. McNeil, Lawrence C. Morris, Jr., Lincoln E. Moses, Seymour Perry, Dorothy P. Rice, Herman A. Tyroler, and Donald A. Young. Study Staff: Enriqueta C. Bond, Barbara Filner, Caren Carney, Clifford Goodman, Linda DePugh, Naomi Hudson, and Wallace K. Waterfall.

B56 John C. Bailar III and Frederick Mosteller, editors. *Medical Uses of Statistics.* Waltham, MA: NEJM Books, 1986.

- Introduction. John C. Bailar III and Frederick Mosteller. pp. xxi–xxvi.
- Chapter 8. James H. Ware, Frederick Mosteller, and Joseph A. Ingelfinger. "P values." pp. 149–169.
- Chapter 13. Rebecca DerSimonian, L. Joseph Charette, Bucknam McPeek, and Frederick Mosteller. "Reporting on methods in clinical trials." pp. 272–288.

  Published earlier—paper P139.

- Chapter 15. Frederick Mosteller. "Writing about numbers." pp. 305–321.

  Italian edition: *L'Uso della Statistica in Medicina.* Roma: Il Pensiero Scientifico Editore, 1988. Translated by Giovanni Apolone, Antonio Nicolucci, Raldano Fossati, Fabio Parazzini, Walter Torri, Roberto Grilli.

PAPERS

P1 Frederick Mosteller. "Note on an application of runs to quality control charts." *Annals of Mathematical Statistics*, **12** (1941), pp. 228–232.

P1a Frederick Mosteller and Philip J. McCarthy. "Estimating population proportions." *Public Opinion Quarterly*, **6** (1942), pp. 452–458.

P2 Louis H. Bean, Frederick Mosteller, and Frederick Williams. "Nationalities and 1944." *Public Opinion Quarterly*, **8** (1944), pp. 368–375.

P3 M.A. Girshick, Frederick Mosteller, and L.J. Savage. "Unbiased estimates for certain binomial sampling problems with applications." *Annals of Mathematical Statistics*, **17** (1946), pp. 13–23.

Reprinted in *The Writings of Leonard Jimmie Savage—A Memorial Selection*, published by The American Statistical Association and The Institute of Mathematical Statistics. Washington, D.C.: The American Statistical Association, 1981. pp. 96–106.

P4 Frederick Mosteller. "On some useful 'inefficient' statistics." *Annals of Mathematical Statistics*, **17** (1946), pp. 377–408.

P5 Cecil Hastings, Jr., Frederick Mosteller, John W. Tukey, and Charles P. Winsor. "Low moments for small samples: a comparative study of order statistics." *Annals of Mathematical Statistics*, **18** (1947), pp. 413–426.

P6 Frederick Mosteller. "A k-sample slippage test for an extreme population." *Annals of Mathematical Statistics*, **19** (1948), pp. 58–65.

P7 Frederick Mosteller. "On pooling data." *Journal of the American Statistical Association*, **43** (1948), pp. 231–242.

P8 Frederick Mosteller and John W. Tukey. "The uses and usefulness of binomial probability paper." *Journal of the American Statistical Association*, **44** (1949), pp. 174–212.

P9 Hendrik Bode, Frederick Mosteller, John Tukey, and Charles Winsor. "The education of a scientific generalist." *Science*, **109** (June 3, 1949), pp. 553–558.

Reprinted in *The Collected Works of John W. Tukey, Volume III, Philosophy and Principles of Data Analysis: 1949–1964*, edited by Lyle V. Jones. Monterey, CA: Wadsworth & Brooks/Cole, 1986. pp. 1–13.

P13 Frederick Mosteller and John W. Tukey. "Practical applications of new theory, a review:"

P10 "Part I: Location and scale: tables." *Industrial Quality Control*, **6**, No. 2 (1949), pp. 5–8.

P11 "Part II. Counted data—graphical methods." *Industrial Quality Control*, **6**, No. 3 (1949), pp. 5–7.

P12 "Part III. Analytical techniques." *Industrial Quality Control*, **6**, No. 4 (1950), pp. 5–8.

P13 "Part IV. Gathering information." *Industrial Quality Control*, **6**, No. 5 (1950), pp. 5–7.

P14 Article on "Statistics." In *Collier's Encyclopedia*, circa 1949, pp. 191–195.

P15 Frederick Mosteller and John W. Tukey. "Significance levels for a k-sample slippage test." *Annals of Mathematical Statistics*, **21** (1950), pp. 120–123.

P16 J.S. Bruner, L. Postman, and F. Mosteller. "A note on the measurement of reversals of perspective." *Psychometrika*, **15** (1950), pp. 63–72.

P17 Arthur S. Keats, Henry K. Beecher, and Frederick Mosteller. "Measurement of pathological pain in distinction to experimental pain." *Journal of Applied Physiology*, **3** (1950), pp. 35–44.

P18 Frederick Mosteller. "Remarks on the method of paired comparisons: I. The least squares solution assuming equal standard deviations and equal correlations." *Psychometrika*, **16** (1951), pp. 3–9.

Reprinted in *Readings in Mathematical Psychology*, **I**, edited by R. Duncan Luce, Robert R. Bush, and Eugene Galanter. New York: Wiley, 1963. pp. 152–158.

P19 Frederick Mosteller. "Remarks on the method of paired comparisons: II. The effect of an aberrant standard deviation when equal standard deviations and equal correlations are assumed." *Psychometrika*, **16** (1951), pp. 203–206.

P20 Frederick Mosteller. "Remarks on the method of paired comparisons: III. A test of significance for paired comparisons when equal standard deviations and equal correlations are assumed." *Psychometrika*, **16** (1951), pp. 207–218.

P21 Frederick Mosteller. "Mathematical models for behavior theory: a brief report on an interuniversity summer research seminar." Social Science Research Council *Items*, **5** (September 1951), pp. 32–33.

P22 Robert E. Goodnow, Henry K. Beecher, Mary A.B. Brazier, Frederick Mosteller, and Renato Tagiuri. "Physiological performance following a hypnotic dose of a barbiturate." *Journal of Pharmacology and Experimental Therapeutics*, **102** (1951), pp. 55–61.

P23 Frederick Mosteller. "Theoretical backgrounds of the statistical methods: underlying probability model used in making a statistical inference." *Industrial and Engineering Chemistry*, **43** (1951), pp. 1295–1297.

P24 Frederick Mosteller and Philip Nogee. "An experimental measurement of utility." *The Journal of Political Economy*, **59** (1951), pp. 371–404.

Reprinted in the Bobbs-Merrill Reprint Series in the Social Sciences.

P25 Robert R. Bush and Frederick Mosteller. "A mathematical model for simple learning." *Psychological Review*, **58** (1951), pp. 313–323.

Reprinted in *Readings in Mathematical Psychology*, **I**, edited by R. Duncan Luce, Robert R. Bush, and Eugene Galanter. New York: Wiley, 1963. pp. 278–288.

Reprinted in the Bobbs-Merrill Reprint Series in the Social Sciences and also reprinted in P84.

P26 Robert R. Bush and Frederick Mosteller. "A model for stimulus generalization and discrimination." *Psychological Review*, **58** (1951), pp. 413–423.

Reprinted in *Readings in Mathematical Psychology*, **I**, edited by R. Duncan Luce, Robert R. Bush, and Eugene Galanter. New York: Wiley, 1963. pp. 289–299.

P27 Frederick Mosteller. "Clinical studies of analgesic drugs: II. Some statistical problems in measuring the subjective response to drugs." *Biometrics*, **8** (1952), pp. 220–226.

P28 Frederick Mosteller. "The World Series competition." *Journal of the American Statistical Association*, **47** (1952), pp. 355–380.

Translated into French by F.M.-Alfred, É.C. "Utilisation de quelques techniques statistiques au service des parieurs. Présentation et traduction de l'article 'World Series Competition' de Fréderic Mosteller." *Hermès*, Bulletin de la Faculté de Commerce de l'Université Laval, 1953, #7, pp. 50–64, #8, pp. 29–37.

P29 Frederick Mosteller. "Statistical theory and research design." *Annual Review of Psychology*, **4** (1953), pp. 407–434.

P30 Robert R. Bush and Frederick Mosteller. "A stochastic model with applications to learning." *Annals of Mathematical Statistics*, **24** (1953), pp. 559–585.

P31 Henry K. Beecher, Arthur S. Keats, Frederick Mosteller, and Louis Lasagna. "The effectiveness of oral analgesics (morphine, codeine, acetylsalicylic acid) and the problem of placebo 'reactors' and 'non-reactors.' " *Journal of Pharmacology and Experimental Therapeutics*, **109** (1953), pp. 393–400.

P32 Frederick Mosteller. "Comments on 'Models for Learning Theory' ", *Symposium on Psychology of Learning Basic to Military Training Problems*, Panel on Training and Training Devices, Committee on Human Resources Research and Development Board, May 7–8, 1953, pp. 39–42.

P33 William G. Cochran, Frederick Mosteller, and John W. Tukey. "Statistical problems of the Kinsey report." *Journal of the American Statistical Association*, **48** (1953), pp. 673–716.

Appeared later in B4, pp. 1–42.

P34 William G. Cochran, Frederick Mosteller, and John W. Tukey. "Principles of sampling." *Journal of the American Statistical Association*, **49** (1954), pp. 13–35.

Appeared later in B4, pp. 309–331.

Translated into Spanish: "Fundamentos de muestreo." *Estadística*, Journal of the Inter-American Statistical Institute, June 1956, XIV, N. 51, pp. 235–258.

P35 Louis Lasagna, Frederick Mosteller, John M. von Felsinger, and Henry K. Beecher. "A study of the placebo response." *American Journal of Medicine*, **16** (1954), pp. 770–779.

P36 Frederick Mosteller and Robert R. Bush. "Selected quantitative techniques." Chapter 8 in *Handbook of Social Psychology*, edited by Gardner Lindzey. Cambridge, MA: Addison-Wesley, 1954. pp. 289–334.

Reprinted in paperback, circa 1969, together with Chapter 9, "Attitude measurement" by Bert F. Green. It was decided to publish these two chapters as a separate book because they were not reprinted in *The Handbook of Social Psychology, Second Edition*, Vol. I, 1968, Vols. II–V, 1969, edited by Gardner Lindzey and Elliot Aronson.

P37 Robert R. Bush, Frederick Mosteller, and Gerald L. Thompson. "A formal structure for multiple-choice situations." Chapter VIII in *Decision Processes*, edited by R.M. Thrall, C.H. Coombs, and R.L. Davis. New York: Wiley, 1954. pp. 99–126.

P38 Frederick Mosteller. Introduction, IV, "Applications." In *Tables of the Cumulative Binomial Probability Distribution*, by the Staff of Harvard Computation Laboratory. *The Annals of the Computation Laboratory of Harvard University*, **XXXV**. Cambridge, MA: Harvard University Press, 1955. pp. xxxiv–lxi.

P39 Frederick Mosteller. "Stochastic learning models." In *Proceedings of the Third Berkeley Symposium on Mathematical Statistics and Probability*, Volume **V**: *Econometrics, Industrial Research, and Psychometry*, edited by Jerzy Neyman. Berkeley: University of California Press, 1956. pp. 151–167.

P40 Frederick Mosteller. "Statistical problems and their solution." Chapter VI in "The measurement of pain, prototype for the quantitative study of subjective responses," by Henry K. Beecher. *Pharmacological Reviews*, **9** (1957), pp. 103–114.

P44 is an expansion of this paper.

P41 Frederick Mosteller. "Stochastic models for the learning process." *Proceedings of the American Philosophical Society*, **102** (1958), pp. 53–59.

Related to P84.

P42 Frederick Mosteller and D.E. Richmond. "Factorial 1/2: A simple graphical treatment." *American Mathematical Monthly*, **65** (1958), pp. 735–742.

P43 Frederick Mosteller. "The mystery of the missing corpus." *Psychometrika*, **23** (1958), pp. 279–289.

Presidential address delivered to the Psychometric Society, Washington, D.C., September 2, 1958.

P44 Frederick Mosteller. "Statistical problems and their solution." Chapter 4 in *Measurement of Subjective Responses, Quantitative Effects of Drugs*, by Henry K. Beecher, M.D. New York: Oxford University Press, 1959. pp. 73–91.

An expansion of P40.

P45 Maurice Tatsuoka and Frederick Mosteller. "A commuting-operator model." Chapter 12 in *Studies in Mathematical Learning Theory*, edited by Robert R. Bush and William K. Estes. Stanford Mathematical Studies in the Social Sciences, III. Stanford: Stanford University Press, 1959. pp. 228–247.

P46 Robert R. Bush and Frederick Mosteller. "A comparison of eight models." Chapter 15 in *Studies in Mathematical Learning Theory*, edited by Robert R. Bush and William K. Estes. Stanford Mathematical Studies in the Social Sciences, III. Stanford: Stanford University Press, 1959. pp. 293–307.

P47 Frederick Mosteller and Maurice Tatsuoka. "Ultimate choice between two attractive goals: predictions from a model." *Psychometrika*, **25** (1960), pp. 1–17.

Reprinted in *Readings in Mathematical Psychology*, **I**, edited by R. Duncan Luce, Robert R. Bush, and Eugene Galanter. New York: Wiley, 1963. pp. 498–514.

P48 Richard Cohn, Frederick Mosteller, John W. Pratt, and Maurice Tatsuoka. "Maximizing the probability that adjacent order statistics of samples from several populations form overlapping intervals." *Annals of Mathematical Statistics*, **31** (1960), pp. 1095–1104.

P49 Frederick Mosteller. "Optimal length of play for a binomial game." *The Mathematics Teacher*, **54** (1961), pp. 411–412.

P50 Frederick Mosteller and Cleo Youtz. "Tables of the Freeman-Tukey transformations for the binomial and Poisson distributions." *Biometrika*, **48** (1961), pp. 433–440.

P51 Gordon W. Allport, Paul H. Buck, Frederick Mosteller, and Talcott Parsons, Chairman. "Samuel Andrew Stouffer," Memorial Minute adopted by the Faculty of Arts and Sciences, Harvard University. *Harvard University Gazette*, April 29, 1961, pp. 197–198.

P52 Frederick Mosteller. "Understanding the birthday problem." *The Mathematics Teacher*, **55** (1962), pp. 322–325.

P53 Frederick Mosteller and David L. Wallace. "Notes on an authorship problem." In *Proceedings of a Harvard Symposium on Digital Computers and Their Applications, 3–6 April 1961.* Cambridge, MA: Harvard University Press, 1962. pp. 163–197.

Report on research later published in B18, B19, and P58.

P54–P57 Reports on Continental Classroom's TV course in probability and statistics.

P54 Frederick Mosteller. "Continental Classroom's TV course in probability and statistics." *The American Statistician*, **16**, No. 5 (December 1962), pp. 20–25.

P55 Frederick Mosteller. "The U.S. Continental Classroom's TV course in probability and statistics." *Quality*, **7**, No. 2 (1963), pp. 36–39.

Abbreviated version.

P56 Frederick Mosteller. "Continental Classroom's TV course in probability and statistics." *The Mathematics Teacher*, **56** (1963), pp. 407–413.

P57 Frederick Mosteller. "Continental Classroom's television course in probability and statistics." *Review of the International Statistical Institute*, **31** (1963), pp. 153–162.

P58 Frederick Mosteller and David L. Wallace. "Inference in an authorship problem: a comparative study of discrimination methods applied to the authorship of the disputed *Federalist* papers." *Journal of the American Statistical Association*, **58** (1963), pp. 275–309.

Report on research later published in B18 and B19; related to P53.

P59 Frederick Mosteller. "Samuel S. Wilks: Statesman of Statistics." *The American Statistician*, **18**, No. 2 (April 1964), pp. 11–17.

P60 Frederick Mosteller. "Contributions of secondary school mathematics to social science." In *Proceedings of the UICSM Conference on the Role of*

Applications in a Secondary School Mathematics Curriculum, edited by Dorothy Friedman. Urbana, Illinois: University of Illinois Committee on School Mathematics, 1964. pp. 85–111.

P61 Frederick Mosteller. "John Davis Williams, 1909–1964, In Memoriam." The Memorial Service, Santa Monica Civic Auditorium, Santa Monica, CA, December 6, 1964, pp. 3–13.

P62 Arthur P. Dempster and Frederick Mosteller. "A model for the weighting of scores." Appendix C in *Staff Leadership in Public Schools: A Sociological Inquiry*, edited by Neal Gross and Robert E. Herriott. New York: Wiley, 1965. pp. 202–216.

P63 Frederick Mosteller. "His writings in applied statistics," pp. 944–953 in "Samuel S. Wilks," by Frederick F. Stephan, John W. Tukey, Frederick Mosteller, Alex M. Mood, Morris H. Hansen, Leslie E. Simon, and W.J. Dixon. *Journal of the American Statistical Association*, **60** (1965), pp. 938–966.

P64 John P. Gilbert and Frederick Mosteller. "Recognizing the maximum of a sequence." *Journal of the American Statistical Association*, **61** (1966), pp. 35–73.

P65 Gene Smith, Lawrence D. Egbert, Robert A. Markowitz, Frederick Mosteller, and Henry K. Beecher. "An experimental pain method sensitive to morphine in man: the submaximum effort tourniquet technique." *Journal of Pharmacology and Experimental Therapeutics*, **154** (1966), pp. 324–332.

P66 Subcommittee on the National Halothane Study of the Committee on Anesthesia, National Academy of Sciences—National Research Council. "Summary of the National Halothane Study: Possible association between halothane anesthesia and postoperative hepatic necrosis." *Journal of the American Medical Association*, **197** (1966), pp. 775–788.

Summary from B24.

P67 Frederick Mosteller. Contribution to "Tribute to Walter Shewhart." *Industrial Quality Control*, **24**, No. 2 (1967), p. 117.

P68 Conrad Taeuber, Frederick Mosteller, and Paul Webbink. "Social Science Research Council Committee on Statistical Training." *The American Statistician*, **21**, No. 5 (December 1967), pp. 10–11.

Reprinted from Social Science Research Council *Items*, **21**, December 1967, pp. 49–51.

P69 Frederick Mosteller. "What has happened to probability in the high school?" *The Mathematics Teacher*, **60** (1967), pp. 824–831.

P87 Frederick Mosteller. "The Jackknife." *Review of the International Statistical Institute*, **39** (1971), pp. 363–368.

P88 Francis J. Anscombe, David H. Blackwell, and Frederick Mosteller (Chairman). "Report of the Evaluation Committee on the University of Chicago Department of Statistics." *The American Statistician*, **25**, No. 3 (June 1971), pp. 17–24.

P89 Frederick Mosteller. "The Joint American Statistical Association–National Council of Teachers of Mathematics Committee on the Curriculum in Statistics and Probability." *Review of the International Statistical Institute*, **39** (1971), pp. 340–342.

P90 Frederick Mosteller. "A data-analytic look at Goldbach counts." *Statistica Neerlandica*, **26** (1972), pp. 227–242.

P91 Frederick Mosteller. "An empirical study of the distribution of primes and litters of primes." Paper 15 in *Statistical Papers in Honor of George W. Snedecor*, edited by T.A. Bancroft and assisted by Susan Alice Brown. Ames, IA: The Iowa State University Press, 1972. pp. 245–257.

P92 William Fairley and Frederick Mosteller. "Trial of an adversary hearing: public policy in weather modification." *International Journal of Mathematical Education in Science and Technology*, **3** (1972), pp. 375–383.

P93 F. Mosteller. "Chairman's introduction." In *Statistics at the School Level*, edited by Lennart Råde. Stockholm, Sweden: Almqvist and Wiksell International. New York: Wiley. 1973, pp. 23–37.

P94 Frederick Mosteller and David C. Hoaglin. "Statistics." In the *Encyclopaedia Britannica*, Fifteenth edition, **17** (1974), pp. 615–624.

P95 Frederick Mosteller. "The role of the Social Science Research Council in the advance of mathematics in the social sciences." Social Science Research Council *Items*, **28**, June 1974, pp. 17–24.

P96 William B. Fairley and Frederick Mosteller. "A conversation about Collins." *The University of Chicago Law Review*, **41**, No. 2 (Winter 1974), pp. 242–253.

Reprinted in B41.

P97 Frederick Mosteller. "Robert R. Bush. Early Career." *Journal of Mathematical Psychology*, **11** (1974), pp. 163–178.

P98 C.F. Mosteller, E.B. Newman, B.F. Skinner, and R.J. Herrnstein, Chairman. "Stanley Smith Stevens," Memorial Minute adopted by the Faculty of Arts and Sciences, Harvard University, April 9, 1974. *Harvard University Gazette*, June 13, 1974.

P99 John P. Gilbert, Richard J. Light, and Frederick Mosteller. "Assessing social innovations: an empirical base for policy." Chapter 2 in *Evaluation and Experiment: Some Critical Issues in Assessing Social Programs*, edited by Carl A. Bennett and Arthur A. Lumsdaine. New York: Academic Press, 1975. pp. 39–193.

This is the original article. Abridgments are P100 and pp. 185–241 in B41.

P100 John P. Gilbert, Richard J. Light, and Frederick Mosteller. "Assessing social innovations: an empirical base for policy." Chapter 1 in *Benefit-Cost and Policy Analysis 1974*, an Aldine Annual on forecasting, decision-making, and evaluation. Edited by Richard Zeckhauser, Arnold C. Harberger, Robert H. Haveman, Laurence E. Lynn, Jr., William A. Niskanen, and Alan Williams. Chicago: Aldine Publishing Company, 1975. pp. 3–65.

An abridgment of P99.

P101 Frederick Mosteller. "Comment by Frederick Mosteller." On David K. Cohen's "The Value of Social Experiments." In *Planned Variation in Education: Should We Give Up or Try Harder?*, edited by Alice M. Rivlin and P. Michael Timpane. Washington, D.C.: The Brookings Institution, 1975. pp. 169–172.

P102 Frederick Mosteller, John P. Gilbert, and Bucknam McPeek. "Measuring the quality of life." In *Surgery in the United States: A Summary Report of the Study on Surgical Services for the United States*, sponsored jointly by The American College of Surgeons and The American Surgical Association, Volume III. Chicago: The College, 1976. pp. 2283–2299.

This original article was slightly revised as Chapter 10 in B40.

P103 John P. Gilbert, Frederick Mosteller, and John W. Tukey. "Steady social progress requires quantitative evaluation to be searching." In *The Evaluation of Social Programs*, edited by Clark C. Abt. Beverly Hills, CA: Sage Publications, 1976. pp. 295–312.

P104 Persi Diaconis, Frederick Mosteller, and Hironari Onishi. "Second-order terms for the variances and covariances of the number of prime factors—including the square free case." *Journal of Number Theory*, **9** (1977), pp. 187–202.

P105 Committee on Community Reactions to the Concorde, Assembly of Behavioral and Social Sciences, National Research Council. "Community reactions to the Concorde: An assessment of the trial period at Dulles Airport." Washington, D.C.: National Academy of Sciences, 1977.

Committee members: Angus Campbell, Chairman, William Baumol, Robert F. Boruch, James A. Davis, Elizabeth A. Deakin, Kenneth M. Eldred, Henning E. von Gierke, Amos H. Hawley, C. Frederick Mosteller, and H. Wayne Rudmose.

P106 John P. Gilbert, Bucknam McPeek, and Frederick Mosteller. "Statistics and ethics in surgery and anesthesia." *Science*, **198** (1977), pp. 684–689.

Reprinted in *Solutions to Ethical and Legal Problems in Social Research*, edited by Robert F. Boruch and Joe S. Cecil. New York: Academic, 1983, pp. 65–82.

P107 Frederick Mosteller. "Experimentation and innovations." *Bulletin of the International Statistical Institute*, Proceedings of the 41st Session, New Delhi, 1977, **XLVII**, Book 1, pp. 559–572.

P108 Frederick Mosteller. "Collegiate football scores, U.S.A." In *Optimal Strategies in Sports*, edited by Shaul P. Ladany and Robert E. Machol. Amsterdam: North-Holland/American Elsevier, 1977. pp. 97–105.

Abridgment of P80.

P109 Frederick Mosteller (with others). "Calendar Year 1977 Report," National Advisory Council on Equality of Educational Opportunity, March 31, 1978. (Includes "Report of the Task Force on Evaluation," pp. 15–24 and "Final Report of the Task Force on Evaluation," pp. 25–38, by Jacquelyne J. Jackson, Haruko Morita, and Frederick Mosteller.)

P110 Oliver Cope, John Hedley-Whyte, Richard J. Kitz, C. Frederick Mosteller, Henning Pontoppidan, William H. Sweet, Leroy D. Vandam, and Myron B. Laver, Chairman. "Henry K. Beecher," Memorial Minute adopted by the Faculty of Medicine, Harvard University, June 1, 1977. *Harvard University Gazette*, January 13, 1978. p. 9.

P111 Frederick Mosteller. "Dilemmas in the concept of unnecessary surgery." *Journal of Surgical Research*, **25** (1978), pp. 185–192.

P112 Frederick Mosteller. "Errors: I. Nonsampling errors." In *International Encyclopedia of Statistics*, edited by William H. Kruskal and Judith M. Tanur. New York: The Free Press, 1978. pp. 208–229.

Slight expansion of P74.

P113 Frederick Mosteller. "A resistant analysis of 1971 and 1972 professional football." In *Sports, Games, and Play: Social and Psychological Viewpoints*, edited by Jeffrey H. Goldstein. Hillsdale, NJ: Lawrence Erlbaum Associates, 1979. pp. 371–399.

114 Frederick Mosteller. Comment on "Field experimentation in weather modification," by Roscoe R. Braham, Jr. *Journal of the American Statistical Association*, **74** (1979), pp. 88–90.

115 Frederick Mosteller. "Problems of omissions in communications." *Clinical Pharmacology and Therapeutics*, **25**, No. 5, Part 2 (1979), pp. 761–764.

116 Frederick Mosteller and Gale Mosteller. "New statistical methods in public policy. Part I: Experimentation." *Journal of Contemporary Business*, **8**, No. 3 (1979), pp. 79–92.

117 Frederick Mosteller and Gale Mosteller. "New statistical methods in public policy. Part II: Exploratory data analysis." *Journal of Contemporary Business*, **8**, No. 3 (1979), pp. 93–115.

118 William Kruskal and Frederick Mosteller. "Representative sampling, I: Non-scientific literature." *International Statistical Review*, **47** (1979), pp. 13–24.

119 William Kruskal and Frederick Mosteller. "Representative sampling, II: Scientific literature, excluding statistics." *International Statistical Review*, **47** (1979), pp. 111–127.

120 William Kruskal and Frederick Mosteller. "Representative sampling, III: The current statistical literature." *International Statistical Review*, **47** (1979), pp. 245–265.

121 William Kruskal and Frederick Mosteller. "Representative sampling, IV: The history of the concept in statistics, 1895–1939." *International Statistical Review*, **48** (1980), pp. 169–195.

122 Frederick Mosteller. "Classroom and platform performance." *The American Statistician*, **34** (1980), pp. 11–17.

123 Frederick Mosteller, Bucknam McPeek, and John P. Gilbert. "The clinician's responsibility for the decision process." *CAHP News Letter* (Center for the Analysis of Health Practices, Harvard School of Public Health), Winter 1980, pp. 2–4.

124 Bucknam McPeek, John P. Gilbert, and Frederick Mosteller. "The clinician's responsibility for helping to improve the treatment of tomorrow's patients." *New England Journal of Medicine*, **302** (1980), pp. 630–631.

125 Clark C. Abt and Frederick Mosteller. "Presentation and acceptance of the Lazarsfeld Prize to Frederick Mosteller." In *Problems in American Social Policy Research*, edited by Clark C. Abt. Cambridge, MA: Abt Books, 1980. pp. 273–276.

Frederick Mosteller's acceptance, pp. 274–276.

P126 Frederick Mosteller, John P. Gilbert, and Bucknam McPeek. "Reporting standards and research strategies for controlled trials: agenda for the editor." *Controlled Clinical Trials*, **1** (1980), pp. 37–58.

P127 F.C. Mosteller. "Clinical trials methodology: hypotheses, designs, and criteria for success or failure." In *Medical Advances through Clinical Trials: A Symposium on Design and Ethics of Human Experimentation, May 31 and June 1, 1979.* Edmonton, Alberta, Canada, edited by John B. Dossetor, 1980. pp. 12–26.

P128 Stephen Lagakos and Frederick Mosteller. "A case study of statistics in the regulatory process: the FD&C Red No. 40 experiments." *Journal of the National Cancer Institute*, **66** (1981), pp. 197–212.

P129 Frederick Mosteller. "Innovation and evaluation." *Science*, **211** (27 February 1981), pp. 881–886.

Presidential address, annual meeting of the American Association for the Advancement of Science in Toronto, Ontario, Canada, 6 January 1981.

P130 Frederick Mosteller. Leonard Jimmie Savage Memorial Service Tribute, Yale University, March 18, 1972. Published in *The Writings of Leonard Jimmie Savage—A Memorial Selection*, edited by a committee. Washington, D.C.: The American Statistical Association and The Institute of Mathematical Statistics, 1981. pp. 25–28.

P131 Arthur P. Dempster and Frederick Mosteller. "In memoriam: William Gemmell Cochran, 1909–1980." *The American Statistician*, **35** (1981), p. 38.

P132 Frederick Mosteller. "Evaluation: requirements for scientific proof." Chapter 8 in *Coping with the Biomedical Literature: A Primer for the Scientist and the Clinician*, edited by Kenneth S. Warren. New York: Praeger, 1981. pp. 103–121.

P133 Frederick Mosteller, Andrew F. Siegel, Edward Trapido, and Cleo Youtz. "Eye fitting straight lines." *The American Statistician*, **35** (1981), pp. 150–152.

Expanded version in B54.

P134 Frederick Mosteller. "Foreword." *Milbank Memorial Fund Quarterly/Health and Society*, **59**, No. 3 (1981), pp. 297–307.

This introduces and discusses a special issue devoted to medical experimentation and social poliy.

P135 William Kruskal and Frederick Mosteller. "Ideas of representative sampling." In *New Directions for Methodology of Social and Behavioral Science: Problems with Language Imprecision*, edited by Donald W. Fiske. San Francisco: Jossey-Bass, 1981. pp. 3–24.

P136 Thomas A. Louis, Frederick Mosteller, and Bucknam McPeek. "Timely topics in statistical methods for clinical trials." *Annual Review of Biophysics and Bioengineering*, **11** (1982), pp. 81–104.

P137 Bucknam McPeek, Cornelia McPeek, and Frederick Mosteller. "In memoriam: John Parker Gilbert (1926–1980)." *The American Statistician*, **36** (1982), p. 37.

P138 Frederick Mosteller. "Foreword." In *Contributions to Statistics: William G. Cochran*, compiled by Betty I.M. Cochran. New York: Wiley, 1982. pp. vii–xiii.

P139 Rebecca DerSimonian, L. Joseph Charette, Bucknam McPeek, and Frederick Mosteller. "Reporting on methods in clinical trials." *New England Journal of Medicine*, **306** (1982), pp. 1332–1337.

Published later in B56. P139 is the first article in "Statistics in Practice," a new series on biostatistics in the *New England Journal of Medicine.*

P140 Frederick Mosteller and John W. Tukey. "Combination of results of stated precision: I. The optimistic case." *Utilitas Mathematica*, **21A** (1982), pp. 155–179.

The first of two special volumes issued on the occasion of the eightieth birthday of Dr. Frank Yates. A sequel to this paper is P149.

P141 Arthur Dempster, Margaret Drolette, Myron Fiering, Nathan Keyfitz, David D. Rutstein, and Frederick Mosteller, Chairman. "William Gemmell Cochran," Memorial Minute adopted by the Faculty of Arts and Sciences, Harvard University, November 9, 1982. *Harvard University Gazette*, December 3, 1982, p. 4.

P142 Frederick Mosteller. "The role of statistics in medical research." Chapter 1 in *Statistics in Medical Research: Methods and Issues, with Applications in Cancer Research*, edited by Valerie Miké and Kenneth E. Stanley. New York: Wiley, 1983. pp. 3–20.

P143 J.L. Hodges, Jr., Frederick Mosteller, and Cleo Youtz. "Allocating loss of precision in the sample mean to wrong weights and redundancy in sampling with replacement from a finite population." In *A Festschrift for Erich L. Lehmann in Honor of His Sixty-Fifth Birthday*, edited by Peter J. Bickel, Kjell A. Doksum, and J.L. Hodges, Jr. Belmont, CA: Wadsworth, 1983. pp. 239–248.

P144 Frederick Mosteller. "The changing role of the statistician: getting into the mainstream with policy decisions." *Section Newsletter*, Statistics, American Public Health Association, February 1983.

Talk given at luncheon in honor of Joel Kleinman, the Spiegelman Award winner, American Public Health Association Annual Meeting, Montreal, Quebec, Canada, November 16, 1982.

P145 Frederick Mosteller. Comment on "Ethical Guidelines for Statistical Practice: Historical Perspective, Report of the ASA Ad Hoc Committee on Professional Ethics, and Discussion." *The American Statistician*, **37** (1983), pp. 10–11.

P146 Frederick Mosteller, John P. Gilbert, and Bucknam McPeek. "Controversies in design and analysis of clinical trials." In *Clinical Trials: Issues and Approaches*, edited by Stanley H. Shapiro and Thomas A. Louis. New York: Marcel Dekker, 1983. pp. 13–64.

P147 John D. Emerson, Bucknam McPeek, and Frederick Mosteller. "Reporting clinical trials in general surgical journals." *Surgery*, **95** (1984), pp. 572–579.

P148 Frederick Mosteller. "Biography of John W. Tukey." In *The Collected Works of John W. Tukey, Volume I, Time Series: 1949–1964*, edited by David R. Brillinger. Belmont, CA: Wadsworth Advanced Books and Software, 1984. pp. xv–xvii.

Included in each volume of *The Collected Works of John W. Tukey.*

P149 Frederick Mosteller and John W. Tukey. "Combination of results of stated precision: II. A more realistic case." In *W.G. Cochran's Impact on Statistics*, edited by Poduri S.R.S. Rao and Joseph Sedransk. New York: Wiley, 1984. pp. 223–252.

Sequel to P140.

P150 Frederick Mosteller. "Selection of papers by quality of design, analysis, and reporting." Chapter 6 in *Selectivity in Information Systems: Survival of the Fittest*, edited by Kenneth S. Warren. New York: Praeger, 1985. pp. 98–116.

P151 Thomas A. Louis, Harvey V. Fineberg, and Frederick Mosteller. "Findings for public health from meta-analysis." *Annual Review of Public Health*, **6** (1985), pp. 1–20.

P152 Frederick Mosteller and Milton C. Weinstein. "Toward evaluating the cost-effectiveness of medical and social experiments." Chapter 6 in *Social Experimentation*, edited by Jerry A. Hausman and David A. Wise. Chicago: University of Chicago Press, 1985. pp. 221–249.

153 Frederick Mosteller and David C. Hoaglin. "Description or prediction?" *1984 Proceedings of the Business and Economic Statistics Section.* Washington, D.C.: American Statistical Association, 1985. pp. 11–15.

154 Judy O'Young, Bucknam McPeek, and Frederick Mosteller. "The clinician's role in developing measures for quality of life in cardiovascular disease." *Quality of Life and Cardiovascular Care*, **1** (1985), pp. 290–296.

155 B. McPeek, M. Gasko, and F. Mosteller. "Measuring outcome from anesthesia and operation." *Theoretical Surgery*, **1** (1986), pp. 2–9.

156 Augustine Kong, G. Octo Barnett, Frederick Mosteller, and Cleo Youtz. "How medical professionals evaluate expressions of probability." *New England Journal of Medicine*, **315** (1986), pp. 740–744.

Response to comments: *New England Journal of Medicine*, **316** (1987), p. 551. [Comments: pp. 549–551.]

157 Stephen W. Lagakos and Frederick Mosteller. "Assigned shares in compensation for radiation-related cancers." *Risk Analysis*, **6** (1986), pp. 345–357.

Response to comments, pp. 377–380. [Comments: pp. 363–375.] Related to B52 and B53.

158 M.F. McKneally, D.S. Mulder, A. Nachemson, F. Mosteller, and B. McPeek. "Facilitating scholarship: creating the atmosphere, setting, and teamwork for research." Chapter 5 in *Principles and Practice of Research: Strategies for Surgical Investigators*, edited by Hans Troidl, Walter O. Spitzer, Bucknam McPeek, David S. Mulder, and Martin F. McKneally. New York: Springer-Verlag, 1986. pp. 36–42.

159 Frederick Mosteller. "A statistical study of the writing styles of the authors of *The Federalist* papers." *Proceedings of the American Philosophical Society*, **131** (1987), pp. 132–140.

Related to B18, B19, P53, and P58.

160 Morris Hansen and Frederick Mosteller. "William Gemmell Cochran, July 15, 1909–March 29, 1980." In *Biographical Memoirs*, **56**. Washington, D.C.: National Academy Press, 1987. pp. 60–89.

161 Kathryn Lasch, Alesia Maltz, Frederick Mosteller, and Tor Tosteson. "A protocol approach to assessing medical technologies." *International Journal of Technology Assessment in Health Care*, **3** (1987), pp. 103–122.

162 Frederick Mosteller. "Implications of measures of quality of life for policy development." *Journal of Chronic Diseases*, **40** (1987), pp. 645–650.

P163 Frederick Mosteller. "Assessing quality of institutional care." *American Journal of Public Health*, **77** (1987), pp. 1155–1156.

P164 Frederick Mosteller. "Compensating for radiation-related cancers by probability of causation or assigned shares." *Bulletin of the International Statistical Institute*, Proceedings of the 46th Session, Tokyo, 1987, **LII**, Book 4, pp. 571–577.

P165 W.H. Kruskal and F. Mosteller. "Representative sampling." In *Encyclopedia of Statistical Sciences*, **8**, edited by Samuel Kotz and Norman L. Johnson. New York: Wiley, 1988. pp. 77–81.

P166 John C. Bailar III and Frederick Mosteller. "Guidelines for statistical reporting in articles for medical journals: amplifications and explanations." *Annals of Internal Medicine*, **108** (1988), pp. 266–273.

P167 Frederick Mosteller. "Broadening the scope of statistics and statistical education." *The American Statistician*, **42** (1988), pp. 93–99.

Pfizer Colloquium Lecture, Storrs, Connecticut, May 11, 1987.

P168 Frederick Mosteller and John W. Tukey. "A conversation." *Statistical Science*, **3** (1988), pp. 136–144.

P169 Graham A. Colditz, James N. Miller, and Frederick Mosteller. "The effect of study design on gain in evaluations of new treatments in medicine and surgery." *Drug Information Journal*, **22** (1988), pp. 343–352.

P170 Graham A. Colditz, James N. Miller, and Frederick Mosteller. "Measuring gain in the evaluation of medical technology." *International Journal of Technology Assessment in Health Care*, **4** (1988), pp. 637–642.

P171 Nan Laird and Frederick Mosteller. Discussion of "Publication bias: a problem in interpreting medical data," by Colin B. Begg and Jesse A. Berlin. *Journal of the Royal Statistical Society, Series A*, **151** (1988), p. 456.

P172 Frederick Mosteller. "Growth and advances in statistics." Response to "Should mathematicians teach statistics?" by David S. Moore. *College Mathematics Journal*, **19** (1988), pp. 15–16.

P173 Frederick Mosteller. " 'The muddiest point in the lecture' as a feedback device." *On Teaching and Learning: The Journal of the Harvard-Danforth Center*, **3** (April 1989), pp. 10–21.

P174 Frederick Mosteller and Elisabeth Burdick. "Current issues in health care technology assessment." *International Journal of Technology Assessment in Health Care*, **5** (1989), pp. 123–136.

P175 The LORAN Commission. "The LORAN Commission: A summary report." In *Harvard Community Health Plan, 1988 Annual Report*, pp. 3–6, 9–14, 17–22, 25–30. (Published in 1989)

Members of the Commission: David Banta, Robert Cushman, Douglas Fraser, Robert Freeman, Betty Friedan, Benjamin Kaplan, Frederick Mosteller, David Nathan, Albert Rees, Hays Rockwell, Robert Sproull, Marshall Wolf, and John Paris.

P176 Frederick Mosteller, John E. Ware, Jr., and Sol Levine. "Finale panel: comments on the conference on advances in health status assessment." *Medical Care*, **27**, No. 3 Supplement (1989), pp. S282–S294.

P177 Graham A. Colditz, James N. Miller, and Frederick Mosteller. "How study design affects outcomes in comparisons of therapy. I: medical." *Statistics in Medicine*, **8** (1989), pp. 441–454.

P178 James N. Miller, Graham A. Colditz, and Frederick Mosteller. "How study design affects outcomes in comparisons of therapy. II: surgical." *Statistics in Medicine*, **8** (1989), pp. 455–466.

P179 Frederick Mosteller. "Summary remarks." *Proceedings of the Workshop on the Future of Meta-Analysis*, Hedgesville, WV, October 1986. In press.

P180 Robert Timothy Reagan, Frederick Mosteller, and Cleo Youtz. "The quantitative meanings of verbal probability expressions." *Journal of Applied Psychology*, **74** (1989), pp. 433–442.

P181 Persi Diaconis and Frederick Mosteller. "Methods for studying coincidences." *Journal of the American Statistical Association*, **84** (1989), pp. 853–861.

P182 Frederick Mosteller and Cleo Youtz. "Quantifying probabilistic expressions." *Statistical Science*. In press.

MISCELLANEOUS

M1—M5 Articles on magic.

M1 Frederick Mosteller. "Encore," Part I. In *My Best*, edited by J.G. Thompson, Jr. Philadelphia: Charles H. Hopkins & Co., 1945. pp. 103–104.

Related to M3.

M2—M5 Appeared in *The Phoenix*, a two-sheet biweekly publication. Issues 1 through 73 were edited by Walter Gibson and Bruce Elliott; later issues were edited by Bruce Elliott. All issues were published by Louis Tannen.

Issues have now been bound into sets of 50 (1–50, 50–100, etc.) and distributed by Louis Tannen Inc., New York, NY.

M2 "Bravo." Issue 49, December 3, 1943, pp. 200–201.

M3 "Encore." Issue 58, April 14, 1944, pp. 236–237.

M4 "Ambiguous." Issue 117, January 10, 1947, p. 470.

M5 "Thesis." Issue 118, January 24, 1947, p. 475.

M6 and M7 are in "Letters from Readers," in *The Bridge World*, edited by Ely Culbertson et al.:

M6 Frederick Mosteller. "Anti Zankl." **17**, No. 8, May 1946, pp. 2–3.

M7 Frederick Mosteller. "Eight points." **17**, No. 12, September 1946, p. 2.

M8 Frederick Mosteller. Contribution to *Standard Sampling Procedures.* Material Inspection Service, U.S.N. Administration Manual, 1945.

M8a Frederick Mosteller and John W. Tukey. Binomial probability graph paper (No. 32,298). Norwood, MA: Codex Book Company, 1946.

M9 Frederick Mosteller. Editor of "Questions and Answers" in *The American Statistician*, October 1947–December 1951. Questions 1–30.

M10 Frederick Mosteller. "Can you be a successful gambler?" *TV Guide*, May 13–19, 1961, pp. 6–7.

M11 Frederick Mosteller. "Textbook supplements." In Gottfried E. Noether, *Guide to Probability and Statistics*, especially prepared for Continental Classroom. Reading, MA: Addison-Wesley, 1961. pp. 43–51.

M12 Frederick Mosteller. "Foreword." In *Probability and Statistics—An Introduction Through Experiments*, by Edmund C. Berkeley. New York: Science Materials Center, 1961. pp. v–vii.

M13 *Goals for School Mathematics.* The Report of the Cambridge Conference on School Mathematics. Educational Services Incorporated. Boston: Houghton Mifflin, 1963.

Frederick Mosteller, contributor.

M14 Frederick Mosteller. "Foreword." In *Math and After Math*, by Robert Hooke and Douglas Shaffer. New York: Walker and Company, 1965. pp. ix–xii.

M15 Frederick Mosteller. "The President Reports: Three Major ASA Actions." *The American Statistician*, **21** (4), 1967, pp. 2–4.

M16 Coeditor of Reports of the National Assessment of Educational Progress. A Project of the Education Commission of the States.

Report 2. *Citizenship: National Results.* November 1970. Washington, D.C.: Superintendent of Documents, U.S. Government Printing Office.

Report 3. *1969–1970 Writing: National Results.* November 1970. Washington, D.C.: Superintendent of Documents, U.S. Government Printing Office.

Report 2-1. *Citizenship: National Results.* November 1970. Education Commission of the States, Denver, Colorado and Ann Arbor, Michigan.

Report 4. *1969–1970 Science: Group Results for Sex, Region, and Size of Community.* April 1971. Washington, D.C.: Superintendent of Documents, U.S. Government Printing Office.

Report 5. *1969–1970 Writing: Group Results for Sex, Region, and Size of Community* (Preliminary Report). April 1971. Education Commission of the States, Denver, Colorado and Ann Arbor, Michigan.

Report 7. *1969–1970 Science: Group and Balanced Group Results for Color, Parental Education, Size and Type of Community and Balanced Group Results for Region of the Country, Sex.* December 1971. Education Commission of the States, Denver, Colorado.

Report 8. *National Results—Writing Mechanics.* February 1972. Washington, D.C.: Superintendent of Documents, U.S. Government Printing Office.

Report 9. *Citizenship: 1969–1970 Assessment: Group Results for Parental Education, Color, Size and Type of Community.* May 1972. Education Commission of the States, Denver, Colorado.

Report 02-GIY. *Reading and Literature: General Information Yearbook.* May 1972. Education Commission of the States, Denver, Colorado.

M17 Frederick Mosteller. "Introduction." In *Structural and Statistical Problems for a Class of Stochastic Processes, The First Samuel Stanley Wilks Lecture at Princeton University, March 17, 1970, by Harald Cramér.* Princeton, NJ: Princeton University Press, 1971. pp. 1–2.

M18 Frederick Mosteller. A page on subject matter at secondary level. In *Developments in Mathematical Education, Proceedings of the Second International Congress on Mathematical Education*, edited by A.G. Howson. Cambridge: The University Press, 1973. p. 27.

M19 Frederick Mosteller. "Foreword." In *Social Statistics in Use*, by Philip M. Hauser. New York: Russell Sage Foundation, 1975. pp. vii–viii.

M20 Frederick Mosteller. "Report of the President." In "Officers' Reports, 1975." *The Institute of Mathematical Statistics Bulletin*, **4** (1975), pp. 207–208.

M21 Frederick Mosteller. Research Resources Evaluation Panel, coordinated by Bolt, Beranek, and Newman. *Assuring the Resources for Biomedical Research: an Evaluation of the Scientific Mission of the Division of Research Resources*, National Institutes of Health. October 1976.

M22 Frederick Mosteller. "Swine Flu: Quantifying the 'Possibility.' " Letter in *Science*, **192** (25 June 1976), pp. 1286 and 1288.

M23 Frederick Mosteller. "Who Said It?" In *Royal Statistical Society News & Notes*, **5**, No. 2, October 1978.

M24 Frederick Mosteller. Testimony by Frederick Mosteller. In "Kenneth Prewitt, Frederick Mosteller, and Herbert A. Simon testify at National Science Foundation Hearings." Social Science Research Council *Items*, **34** (March 1980), pp. 4–5.

All testimonies pp. 1–7.

M25 Frederick Mosteller. "Regulation of social research." Editorial in *Science*, **208** (13 June 1980), p. 1219.

M26 Frederick Mosteller. "The next 100 years of *Science*." In "Science Centennial 3 July 1980 to 4 July 1980," Centennial Issue of *Science*, edited by Philip H. Abelson and Ruth Kulstad. **209** (4 July 1980), pp. 21–23.

M27 Frederick Mosteller. "Social programs." *Transaction/Social Science and Modern Society*. **17**, No. 6 (September/October 1980), pp. 10–12.

M28 Frederick Mosteller. "Taking science out of social science." Editorial in *Science*, **212** (17 April 1981), p. 291.

M29 Frederick Mosteller. "Improving the precision of clinical trials." Editorial in *American Journal of Public Health*, **72** (May 1982), p. 430.

M30 Frederick Mosteller. "The imperfect science of victim compensation." Washington, D.C.: *The Washington Times*, June 4, 1985. p. 12A.

M31 Frederick Mosteller. Abstract of talk on "Assigned Shares: probability of radiation as a cause of cancer." Final Report, ASA Conference on Radiation and Health, Coolfont V. Washington, D.C.: American Statistical Association, 1985. pp. 21–22.

M32 and M33 are in *Data. A Collection of Problems from Many Fields for the Student and Research Worker*, edited by D.F. Andrews and A.M. Herzberg. New York: Springer-Verlag, 1985.

M32 S.W. Lagakos and F. Mosteller. "Time to death and type of death in mice receiving various doses of Red Dye No. 40." pp. 239–243.

Data from P128.

M33 F. Mosteller and D.L. Wallace. "Disputed authorship: *The Federalist* Papers." pp. 423–425.

Data from B18.

M34 Frederick Mosteller. "Foreword." In *New Developments in Statistics for Psychology and the Social Sciences*, edited by A.D. Lovie. London and New York: The British Psychological Society and Methuen, 1986. pp. vii–ix.

M35 Barbara J. Culliton and Frederick Mosteller. "How big is 'big,' How rare is 'rare?' Inquiring minds want to know." *The Newsletter of the National Association of Science Writers*, **34**, No. 3 (September 1986), p. 13.

M36 Frederick Mosteller. "Foreword." In *News & Numbers: A Guide to Reporting Statistical Claims and Controversies in Health and Related Fields*, by Victor Cohn. Ames, IA: Iowa State University Press, 1989. pp. ix–x.

REVIEWS

R1 *Guide for Quality Control and Control Chart Method of Analyzing Data.* American War Standards, Z1.1–1941 and Z1.2–1941. New York: American Standards Association, 1941, and *Control Chart Method of Controlling Quality During Production.* American War Standards, Z1.3–1942. New York: American Standards Association, 1942.
Frederick Mosteller. *Journal of the American Statistical Association*, **40** (1945), pp. 379–380.

R2 Ledyard R. Tucker. "Maximum validity of a test with equivalent items." *Psychometrika*, **11** (1946), pp. 1–13.
F. Mosteller. *Mathematical Reviews*, **7** (1946), pp. 463–464.

R3 Frederick E. Croxton and Dudley J. Cowden. "Tables to facilitate computation of sampling limits of s, and fiducial limits of sigma." *Industrial Quality Control*, **3** (July 1946), pp. 18–21.
Frederick Mosteller. *Mathematical Tables and Other Aids to Computation*, **II**, No. 18, April 1947. National Research Council. p. 258.

R4 Hans Zeisel (Introduction by Paul F. Lazarsfeld). *Say It with Figures.* New York: Harper and Brothers, 1947.
Frederick Mosteller. *Public Opinion Quarterly,* **11** (Fall 1947), pp. 468–469.

R5 Quinn McNemar. "Opinion-attitude methodology." *Psychological Bulletin,* **43**, No. 4 (July 1946), pp. 289–374. Washington, D.C.: American Psychological Association.
Frederick Mosteller. *Journal of the American Statistical Association,* **42** (1947), pp. 192–195.

R6 Paul G. Hoel. *Introduction to Mathematical Statistics.* New York: Wiley, 1947.
Frederick Mosteller. *The Journal of Business of the University of Chicago,* **20** (1947), pp. 176–177.

R7 George W. Snedecor. *Statistical Methods.* Ames, Iowa. The Iowa State College Press, 1946.
Frederick Mosteller. *Annals of Mathematical Statistics,* **19** (1948), pp. 124–126.

R8 Abraham Wald. *Sequential Analysis.* New York: Wiley, 1947.
Frederick Mosteller. *Journal of Applied Mechanics,* **15** (1948), pp. 89–90.

R9 Palmer O. Johnson. *Statistical Methods in Research.* New York: Prentice-Hall, 1949.
Frederick Mosteller. *Journal of the American Statistical Association,* **44** (1949), pp. 570–572.

R10 Norbert Wiener. *Cybernetics, or Control and Communication in the Animal and the Machine.* New York: Wiley, 1948.
Frederick Mosteller. *Journal of Abnormal and Social Psychology,* **44** (1949), pp. 558–560.

R11 N. Rashevsky. *Mathematical Theory of Human Relations: An Approach to a Mathematical Biology of Social Phenomena.* Mathematical Biophysics Monograph Series No. 2. Bloomington, IN: Principia Press, 1948.
Frederick Mosteller. *Journal of the American Statistical Association,* **44** (1949), pp. 150–155.

R12 S.S. Wilks. *Elementary Statistical Analysis.* Princeton: Princeton University Press, 1948.
Frederick Mosteller. *Psychometrika,* **15** (1950), pp. 73–76.

R13 Frank Yates. *Sampling Methods for Censuses and Surveys.* London: Charles Griffin and Company, 1949.

Frederick Mosteller. *The Review of Economics and Statistics*, **32** (1950), pp. 267–268.

R14 W.E. Deming. *Some Theory of Sampling*. New York: Wiley, 1950.
Frederick Mosteller. *Psychological Bulletin*, **48** (1951), pp. 454–455.

R15 *Gamma Globulin in the Prophylaxis of Poliomyelitis: An evaluation of the efficacy of gamma globulin in the prophylaxis of paralytic poliomyelitis as used in the United States 1953*. Public Health Monograph No. 20. Report of the National Advisory Committee for the Evaluation of Gamma Globulin in the Prophylaxis of Poliomyelitis, Public Health Publication No. 358, U.S. Department of Health, Education, and Welfare. Washington, D.C.: Superintendent of Documents, U.S. Government Printing Office, 1954.
Frederick Mosteller. *Journal of the American Statistical Association*, **49** (1954), pp. 926–927.

R16 Paul F. Lazarsfeld, editor. *Mathematical Thinking in the Social Sciences*. Glencoe, Illinois: The Free Press, 1954.
Frederick Mosteller. *American Anthropologist*, **58** (1956), pp. 736–739.

R17 John Cohen and Mark Hansel. *Risk and Gambling: The Study of Subjective Probability*. London: Longmans, Green, and Co., 1956.
Frederick Mosteller. *Econometrica*, **27** (1959), pp. 505–506.

R18 Herbert A. Simon. *Models of Man: Social and Rational. Mathematical Essays on Rational Human Behavior in a Social Setting*. New York: Wiley, 1957.
Frederick Mosteller. *American Sociological Review*, **24** (1959), pp. 409–413.

Contains some original data illustrating the changing values of a parameter of a stochastic model—the probability of a new word—as the number of words in the text so far increases.

R19 Raoul Naroll. *Data Quality Control—A New Research Technique. Prolegomena to a Cross-Cultural Study of Culture Stress*. New York: The Free Press, 1962.
Frederick Mosteller and E.A. Hammel.
Journal of the American Statistical Association, **58** (1963), pp. 835–836.

R20 Herbert Solomon, editor. *Studies in Item Analysis and Prediction*. Stanford Mathematical Studies in the Social Sciences, VI. Stanford, CA: Stanford University Press, 1961.
Frederick Mosteller. *Journal of the American Statistical Association*, **58** (1963), pp. 1180–1181.

R21 L. Råde, editor. *The Teaching of Probability and Statistics*. Stockholm: Almqvist & Wiksell, 1970. Proceedings of the first CSMP International

Conference co-sponsored by Southern Illinois University and Central Midwestern Regional Educational Laboratory.
F. Mosteller. *Review of the International Statistical Institute*, **39**, No. 3 (1971), pp. 407–408.

R22 Frederick Mosteller, Gale Mosteller, and Keith A. Soper. "Knowledge beyond achievement. A seventies perspective on school effects—a review symposium on *The Enduring Effects of Education*, by Herbert Hyman, Charles Wright, and John Reed." *School Review*, **84** (1976), pp. 265–283.

R23 Anthony C. Atkinson and Stephen E. Fienberg, editors. *A Celebration of Statistics (The ISI Centenary Volume)*. New York: Springer-Verlag, 1985.
Frederick Mosteller. *Journal of the American Statistical Association*, **81** (1986), pp. 1118–1119.

R24 Hans Zeisel. *Say It with Figures*, Sixth Edition. New York: Harper & Row, 1985.
Frederick Mosteller. *Public Opinion Quarterly*, **52** (1988), pp. 274–275.

R25 W.S. Peters. *Counting for Something: Statistical Principles and Personalities*. New York: Springer-Verlag, 1987.
Frederick Mosteller. *Metrika*, **36** (1989), pp. 61–62.

Chapter 3

CONTRIBUTIONS AS A SCIENTIFIC GENERALIST

William Kruskal

3.1 Background

Frederick Mosteller and three friends published in 1949 their suggestions toward the education of a scientific generalist, together with their views about the importance of having scientific generalists (P9). The article was about the *undergraduate* education of a budding generalist, yet I regard the exposition as something of a synecdoche for the entire education of a scientific generalist, together with the whole sweep of scientific activities during the generalist's career.

Statisticians and mathematicians, by the intrinsic nature of their fields, tend to be generalists in one sense or another. Indeed, of the four authors of the 1949 article, three are in my view statisticians (John Tukey, Charles Winsor, and Frederick Mosteller), and the fourth (Henrik Bode) an applied mathematician. The four authors, however, made major contributions to many other fields of science; I can list quickly engineering, chemistry, biology, medicine, psychology, sociology, meteorology, and on and on. The complementary list of fields not seriously touched by at least one of the four would be shorter.

I shall first summarize the 1949 article in terms including, but broader than, preparatory education. Then I shall present a classification—inspired by the article—of what functions generalists fulfill. To put flesh on that outline, I will then turn to a nonrandom sample of Fred Mosteller's contributions and show how they fit under the rubrics of the proposed taxonomy.

What then is a scientific generalist, and what does a scientific generalist do? What distinguishes a generalist from other scientists, who, after all, are broadly encouraged to look about and avoid parochialism?

A scientific generalist is, according to the 1949 article, in the first place trained and active in many fields of science. The training is in part formal, but in large part auto-didactic. The generalist has great breadth of appreciation and interest within science, maintains a sense of scientific unity, and handles with aplomb the inevitable tensions between complexity and desired simplicity. Scientific generalists inhabit most parts of the scientific mansion, but they are especially noticeable as university professors, research team leaders, administrators (and particularly advisors to high administrators) of various other kinds, and statisticians. Indeed statisticians and applied mathematicians are generalists almost by the nature of their disciplines, but generalists can be found elsewhere, often under titles or nominal roles that are misleading.

The 1949 article mentions briefly (p. 554), but does not develop, the concept of a trans-science generalist, a transcendent person who understands many fields of science and also many other domains. I sense views half-suggested here about intellectual aristocracy, Platonic views, that may not fit well with other approaches to self-government of a polity. So I regret that the four authors did not give us their opinions on these fundamental issues in 1949 nor, so far as I know, at later times. The issues go back to *The Federalist* papers in this country, and are current in debates about the distribution of support to education and to science. (Fred Mosteller, I note in passing, is a renowned expert on *The Federalist*; the book on that subject (B18, B19)—jointly done with my colleague David L. Wallace—will be described later.)

Among the activities of scientific generalists (and I suppose of generalists transcendent or not) are attempts to state poorly-defined problems, to sharpen them, and to work on them. That a problem does not lie in a single traditional discipline does not discourage a generalist, but typically heightens motivation. Generalists design broad studies and interpret broad data. Of special importance to a generalist is judgment in the sense of intelligent guessing, and the four authors suggest a course in such intelligent guessing during the senior year of a scientific generalist's college training ... intelligent guessing as it closely articulates with scientific method. The main difficulty in arranging for such a course, says the article, "would be finding an instructor equipped to teach it." Mosteller's interest in such order of magnitude estimation continued and found particular expression in a brilliant 1977 article (B41, pp. 163–184).

We have all aged since 1949; many of the readers of the present discussion were then not yet born. Yet the encompassing intelligence and style of 1949 led to an article that has aged little. Details have changed, and I am sure that in 1986 the authors would say some things differently. In essence, however, what they expressed in 1949 seems to me perfectly valid and valuable in 1986. The subsequent scientific activities of the four authors illustrate and produce exemplars for the life of a scientific generalist.

Fred Mosteller provides an exemplary exemplar with his own special flavors. As I think about the sweep of his work, I see the following categories of generalist activities:

1. Creation and generalization within his own field of statistics.

2. Generalization of ideas, theories, developments between statistics and other domains, e.g., psychology and mathematics.

3. Establishing links among non-statistical fields. Translator between, and broker among, separate disciplines.

4. Creator of links between basic scientific research and domains of applications.

5. Formation of new fields or modes of inquiry.

6. Encouragement of scholars—young and old, but especially young—toward exciting and fulfilling work on questions they had not known of. Education more broadly.

7. Institutional leadership in science.

I now turn to illustrate these categories in terms of Fred Mosteller's many contributions.

3.2 Creation and Generalization within Statistics

Three other chapters in this volume deal with Mosteller's research within his own home field of statistics proper. Chapter 4 treats several areas in statistical theory and probability, Chapter 5 discusses statistical methodology and applications, and Chapter 7 focuses on Mosteller's statistical and administrative activities at Harvard University. I add just a few comments to that splendid array of appreciation within statistics.

First ... a personal, perhaps slightly wry, expression of awe. I have had the experience several times of becoming interested in a problem or a domain, working on it, and then finding that Fred Mosteller had been there before me. Here are two examples.

Relative importance. For some time it has seemed to me that issues of relative importance among independent variables, causes, whatever— perhaps "determiners" is a good neutral word—deserve more attention and more serious attention by statisticians. What a pleasure to discover that Fred Mosteller had been there years earlier (B1, Chapter XIV). His interest then was in the relative importance of education and income as they affect responses to attitude questions about important current events. As far as I know, he has not returned to this particular aspect of relative importance, but the Mosteller-Wallace *Federalist* volume (B18, B19) treats two related senses of relative importance of words as they are used in studying authorship ascription (B19, pp. 55–58, 202–204).

Representativeness. That samples represent the population sampled is almost a platitude, yet it is also fundamental. I had from time to time worried about ambiguity in the concept of representativeness, and that worry became a long-term project with Mosteller when we talked about it around

1970 and discovered that both of us had similar concerns. We determined to study the existing ambiguities in concepts of representative sampling, and eventually four papers appeared in the *International Statistical Review* (Kruskal and Mosteller, P118–P121). The first three papers treated, in order, ambiguity of representativeness in non-scientific literature, in scientific literature excluding statistics, and in statistics; the fourth paper was on the history of representative concepts. A fifth paper (P135) summarized the prior four. Our studies led us along many paths: Lenin as a statistician, Fredholm—the mathematician—also as a statistician, Emerson and Wordsworth, some mathematical aspects, and a bevy of illustrative quotations . . . a few delightful, others pompous, and some surprising.

The lessons for me are that, first, Fred Mosteller will probably have been there before in whichever direction my interests turn, and, second, that earlier attention will be stimulating and productive, whether or not we actually carry out joint research.

My second general remark about Fred's research within statistics, although it holds as well for all his research, is that his thinking and his publications form a complex, interwoven pattern, full of life, color, and remarkable interconnections. There are few dangling threads, few roads untravelled.

One example of the Mosteller coherence is his continuing combination of fascination with mathematics and his love of empirical study. For Fred, mathematics has formulation and proof of theorems as its core, but empirical investigation is never far away. An illustration of this combination is his work on the distribution of primes and related matters (P90, P91). A rather different expression of the same intertwining is the paper (with others) on the behavior of dice thrown a great many times (P85).

Another example of continuity and pattern in the Mosteller statistical corpus is his fascination with distribution-free, or nonparametric, methods and their cousins of varying degree: methods based on order statistics. I hasten to add that the terms "distribution-free" and "nonparametric" are not to Fred's liking; he prefers broader and functionally persuasive descriptors, like "sturdy" and " 'inefficient' " (where it is important that the latter word be in quotation marks).

His doctoral work examined the efficacy of "inefficient" statistics, and his first published paper (P1) was on the use of runs in quality control. That topic certainly presaged his enduring concern to bring together theory and practice, and simultaneously to advance both.

Later publications along these lines are on order statistics moments (P5), slippage tests (P6, P15), coverage (P48, P77), and no doubt others. The 1973 book *Sturdy Statistics* (B33a) should be prominently listed here.

These two topics make but a bare start at describing Fred Mosteller's research in statistics proper. The description is much extended elsewhere in this volume.

3.3 Overlaps between Statistics and Other Domains

A scientific generalist surely understands and advances knowledge in the intersection of the generalist's home field—here statistics—with other disciplines. Other chapters in this volume deal with Mosteller's activities in the overlap of statistics with neighboring fields, including psychology, mathematics, medicine, social sciences, and education. Three further Mosteller contributions to important overlaps—law, history, meteorology—will be described next.

Law. Fred Mosteller has long had a deep interest in the use of statistics in the law, and he taught a memorable course at his university's law school. So far as I know, however, he has only one publication in this domain (P96), a discussion—with William B. Fairley—of the infamous Collins case, with its risible estimates of probabilities, and its derisable assumption of independence. The Mosteller theme is that we should pause before splitting our sides in laughter or derision; the Collins assumptions deserve a fairer hearing, and that hearing will help us understand better both law and statistics.

History. The Federalist book (P58, B18, B19), by Mosteller and David L. Wallace, is a landmark in the application of statistics to authorship ascription, and thus—in context—to history, perhaps to constitutional law, and certainly to philosophy ... for it allows a comparison of Bayesian and non-Bayesian methods in a concrete setting of great intrinsic interest.

Meteorology. Four Mosteller publications deal with the complex and contentious field of weather modification. (B38, P79, P92, P114). The problems of design and analysis in this domain, with high variability, unknown mechanisms, and bewildering results, have been worthy of Mosteller's analyses.

One domain, art, has not, to my knowledge, had the advantage of Fred Mosteller's analytic attention, although he has long been deeply interested in art, especially paintings. Perhaps it is essential to each generalist that there be at least one corner of life admired for its own sake only.

3.4 Establishing Links among Nonstatistical Fields

We have already seen examples in which generalist Mosteller combines two or more fields not his own home discipline. For example, the *Federalist* book is relevant to philosophy, history, law, and literature, as well as to statistics.

Public opinion has held much interest for Fred as a composite of sociology, psychology, political science, and—of course—statistics. Some of Mosteller's earliest work was in public opinion, specifically, many parts of Hadley Cantril's 1944 book (B1) and a paper with Bean and Williams on nationalities (P2). The Social Science Research Council published in 1949 its report on the pre-election polls of 1948 (B3), in a multi-author and highly influential volume.

Mosteller's attention extended from opinion polls as such to surveys generally, and one thinks immediately of the wonderful study by him, William Cochran, and John Tukey of the then-hotly discussed Kinsey report on human sexuality (B4). Kinsey's study was based on a sample of convenience and judgment, about as far from a probability sample as one can imagine. His interviewing methods were of highly uncertain validity and reliability. In addition, his findings were shocking to many in the milieu of the day. It would be easy to write an utterly damning analysis, and indeed some were written.

Mosteller and his coauthors took a different approach, trying to find what was defensible in Kinsey's work and going from that specific study to constructive generalizations about other studies meeting similar problems. It was a masterpiece of tolerance and lesson-learning, carried out with great effort, including—I am told—agreement by the three statisticians to Kinsey's insistence that they become part of his files by going through his standard interview.

The Kinsey study was published as a book by the American Statistical Association, and it led to at least two journal articles (P33, P34). Thus we see illustrated another Mosteller characteristic: If the picture is a useful one, paint some more like it, but with variations and for different audiences. Mosteller has consistently expressed a research accomplishment in various ways, not as pedestrian repetition, but rather because he sees a new perspective, or another group of viewers, or a chance to improve the details.

It would be interesting to survey generalists systematically to see whether they are more prone to variant publication than specialists. I do not have a clear guess.

3.5 Links between Research and Applications

Nearly all of Mosteller's scientific work relates to links between basic research and applications. Indeed, it is artificial to make much of a separating line when discussing Mosteller. As I wrote in 1979 (*Science*, **203**, 866–867), "...in the Mosteller cosmos there are no boundaries among different kinds of statistics. For example, the analysis of surgical risks leads to new theoretical developments for cross-classified data; in turn that leads to applications in law and elsewhere."

Another good example is Fred's magisterial paper on non-sampling errors (P112, extending P74), which is expository and encylopedic in the best senses of those words. Basic research results are brought to bear on difficult important problems of measurement errors, systematic errors, etc., and unsolved theoretical issues are outlined.

In psychology, before the famous work on learning theory models, Fred has a superb trio of papers (P18–P20) on paired comparisons, a topic that certainly links basic research with applications. I also note the still earlier paper on vision with Bruner and Postman (P16), and the 1951 paper with Nogee on measuring utility (P24), which appeared in a prestigious economics journal. This joint work with Nogee, a psychologist, was pathbreaking empirical work infused with theoretical understanding. Another paper in the same general domain is with Tatsuoka (P47).

Finally, this may be a reasonable place to cite Fred's long interest in sports and his several papers on statistical analyses or problems in sports. Mostly they are about football (P80; B34, Set 7; B35, Set 11; P108; P113), but one famous paper (P28) is about the World Series in baseball. It was first published in English under the title "The World Series competition" and then in French, but with the different title (back-translating literally) "Use of some statistical methods to help bettors." A minor mystery that I have not delved into. Examples and exercises in Mosteller textbooks also turn to baseball for realistic and motivational context.

3.6 Formation of New Fields of Inquiry

Everything Fred's mind touches is renewed, yet paradoxically nothing is totally new. For all of Fred's creative initiatives one can find precursors ... and that is fitting for work of such deep relevance. I choose to discuss in this section an important facet of Fred's scholarship that could well come under other rubrics (for example, links between basic research and applications) but that in Fred's hands has been transformed into a region of inquiry: statistics as it relates to major questions of public policy. The relationship is, of course, two-way; statistics is applied to public policy questions, their formulation and analysis, but at the same time those public policy questions refresh and motivate statistical thinking. I use the term public policy in a broad sense, including, but not limited to, government policy at all levels. Connections between statistics and public policy are, to be sure, of long standing. One bit of evidence is the etymological overlap of "state" and "statistics." Many of the founding fathers of statistics had vigorous concerns with social issues, for example, Karl Pearson and F. Y. Edgeworth. In Mosteller's hands those concerns have been sharpened, brought to bear in new directions, and made freshly cogent to statistician and non-statistician alike.

Several of these statistics-cum-public policy activities have already been mentioned: public opinion, standards for surveys, relative importance of influences, law, weather modification, and surgical risks. A specifically governmental activity began with Fred's vice-chairmanship of the President's Commission on Federal Statistics (1970–1971) and his energetic leadership—with Chairman W. Allen Wallis—toward the Commission's report (B25, B26). That report described the scope of Federal Statistics by vignettes (Congressional redistricting, testing a polio vaccine, randomization in draft lotteries, price levels, etc.); large achievements of Federal Statistics like the Current Population Survey; opportunities to improve Federal Statistics; privacy and confidentiality; evaluating Federal Statistical programs; and other central topics. Fred (together with Richard J. Light and Herbert S. Winokur, Jr.) has a masterful essay (Chapter 6 of B26) in the Commission's report on the use of controlled field studies, i.e., experiments, to improve public policy. The examples include Head Start, the Martin re-employment study, the New Jersey work incentive experiment, the Baltimore housing study, and smoking and health.

A major recommendation of the President's Commission was establishment of a continuing group, perhaps at the National Academy of Sciences–National Research Council, to undertake studies that could not be done by a one-year ad hoc commission. The recommendation resulted in formation of the Committee on National Statistics at NAS-NRC; Fred and I were among the charter members; his strength of mind and purpose played a large role in getting the Committee on National Statistics off to a vigorous, active start. A partial list of the topics on which the Committee has published studies gives an idea of the breadth of range and relevance as envisioned and encouraged by Fred Mosteller: the national crime survey, statistics as evidence in the courts, monitoring the environment, appraising Census plans, privacy and confidentiality in survey response, measuring productivity, statistics for rural areas, family assistance and poverty, incomplete data, subjective phenomena (e.g., attitudes), immigration statistics, and natural gas data.

Many Mosteller writings deal with statistics in public policy contexts: for example, B25, B26, B41, B44; P99, P100, P101, P103, P116, P117; M25, M27. These, and related studies, have introduced many statisticians to important problems and issues not traditionally found in the academic classroom. For example, realistic approaches to missing data and to selective forces in sampling are only now undergoing wide examination; it is not yet clear how refractory these problems will prove to be. An example of Fred's own theoretical work stemming in part from public policy questions is his ASA presidential address (P73).

I conclude this section with a brief description of one of Fred's most productive initiatives in the public policy domain: his book (B32) with Daniel P. Moynihan (now U. S. Senator from New York), *On Equality of Educational Opportunity.* The book's starting place was the famous study headed by sociologist James S. Coleman with Congressional sponsorship and funds. How unequal are our public schools was a major question of the day, unequal in terms of facilities, teacher ability, etc., and to what extent are those differences reflected in achievement test scores? Special attention was given to region of country and to race. The study was carried out in a great hurry (against a Congressional deadline) and in a climate of political tension. Its analysis and report were stressful and generated wide arguments across the nation, arguments among newspapers, magazines, teachers' groups, government officials, professors happy and angry, etc. These arguments were all the more strident because it appeared that school facility background effects—after allowing for family effects—were amazingly small. The study was criticized and defended for its measuring instruments, its sampling, its tacit assumptions, its statistical analysis, and the lucidity of its report.

Mosteller and Moynihan felt that this unusually large-scale study, with important possible effects for society, deserved a careful hearing and an organized, deliberate debate. They set up a most unusual Harvard seminar, with large contributions from study director Coleman, from his peer academic critics, and from many others interested in various aspects of the study. One such aspect I mentioned much earlier in this essay: how to measure sensibly the relative importance, say of home and school on children's school achievement; how to analyze that further, for example, among school facilities: libraries, laboratories, teachers' training, etc.? How to measure the effect of racial segregation?

The resulting book did not provide resounding univocal answers, nor was it expected to. It did, however, unfold the most important issues with a clarity and concreteness that had great importance.

3.7 Encouragement of Other Scholars

Fred Mosteller has always been a marvel of encouragement to other scholars, young and old ... but with special attention to the young. His talent for bringing others into his manifold research activities is phenomenal, and his encouragement of research by students, colleagues, friends, even strangers, is legendary. One gets the flavor of all this by scanning the many joint authorships in Fred's bibliography. What is harder to examine systematically is the galactic cloud of publications by others for which Fred—formally thanked or not—has been an inspiration, a goad, a friendly nudge in the ribs, a source of just the right technical point at a crucial time.

3.8 Institutional Leadership in Science

Generalists are almost ipso facto bound to take leadership positions in their own home fields and in science generally. Fred Mosteller is an example par excellence. He has served with great distinction on a staggering number of commissions, panels, and boards, often as chairman. He has been president of the American Statistical Association and of the Institute of Mathematical Statistics; on a broader canvas, he served as 1980 president of the American Association for the Advancement of Science. These activities show unflagging energy and a well-deserved reputation for wisdom, discretion, and the ability to integrate and resolve diverse views. William Carey (American Association for the Advancement of Science) prepared for this volume a moving sketch of Fred Mosteller as the Association's president (it forms an appendix to this chapter).

The list of responsible offices and honors, appropriately enough for so model a generalist, spreads its wings in many directions. Fred Mosteller has been (or is) president of the Psychometric Society, chair of two departments at the Harvard School of Public Health (Biostatistics, Health Policy and Management), and Chairman of the Board of Directors of the Social Science Research Council. I stop only because the point is made without trying your patience. Fred is indeed a model model: robust, elegant, teeming with discoveries past and to come, and an example for us all.

Appendix 1: The Mosteller Years at AAAS

William D. Carey

Frederick Mosteller, one of the country's foremost statisticians and scientific statesmen, was a member of the AAAS Board of Directors from 1975 to 1981, and was President of the Association in 1980. He also served on several committees and on the Editorial Boards of *Science* and *Science 85*. He was chairman of Section U (Statistics) in 1972, of the Committee on Council Affairs in 1979, and of the search committee for a new editor of *Science* in 1983–1984.

During those years, Mosteller was a major contributor to policies underlying the whole array of AAAS activities. He regarded *Science* as the Association's flagship, assigned a high priority to the maintenance of its high editorial standards, and took a personal interest in the evaluation of the statistical treatment of scientific material appearing in the journal. His advice to Editor Philip Abelson was influential in the development of the present system of handling statistical matters. Members of the editorial staff were encouraged to sharpen their competence in that field, and the services of a part-time expert consultant were engaged.

Perhaps closest to Mosteller's heart were the Association's programs to improve the elementary and secondary educational process and to promote public understanding of science. From his first days on the Board, he urged that AAAS assume a more active role in these areas.

In an interview published in *Science* shortly before his election as President-Elect, Mosteller identified three targets, besides scientists themselves, at which AAAS should aim its efforts:

> "The first is youth who may well take up science as a career; a second group is young people who ought to be concerned with science because as future citizens they're going to have to deal with science issues more and more. I would like us to bring a 'science and the citizen' program to young people who will have to make important decisions about science as adults."
>
> Adult citizens, the third group, also need a "large-scale educational program" in science. The development of such a program would depend on the cooperation of scientists ... as well as that of "outside leadership"—legislators and policy-makers in cities and towns and in school districts, "people who have to decide on rules and regulations, and who have to debate issues like 'shall there be nuclear facilities,' and 'when shall we permit the use of various kinds of chemicals,' and 'what kinds of steps can be taken to improve health and safety?"
>
> "... AAAS has an obligation to communicate the complexity of science policy to all citizens, and to try to outline options and tradeoffs and provide understanding of the differences in policies. We need to produce information for informed public debates."

Fred Mosteller was an enthusiastic backer of the proposal to sponsor a popular magazine of science that resulted in the successful launching of *Science 80* in late 1979. The current project to provide the nation's public schools with educational materials based on the magazine was an idea that he strongly endorsed.

He was one of the authors of the 1981 resolution in which the AAAS Board and Council pledged the Association, in partnership with its affiliated science and engineering societies, "to a full measure of effort to reverse the damaging decline of science and engineering education in the United States." Fred may view with pride the leading role AAAS has taken in that effort in recent years.

All this says much about Fred's years of leadership of the AAAS, but nothing at all about what he was like to work with. In winter, he invariably arrived at Board meetings dressed in the style of a Maine woodsman and unrecognizable as the famous Harvard professor of biostatistics. His first order of business, in any season, was to investigate the AAAS "book grab," a fringe benefit offered to all members of the Board to take their pick of the accumulated books sent to *Science* by publishers in hopes of a printed review, a boon rarely granted. In his year as chairman, if a contretemps arose among Board members, Fred would let it run on long enough for all views to be civilly heard and then, with mild but deadly elocution, swiftly put a face-saving end to it.

But he was at his best in his dealings with the AAAS staff, most of whom were young and responsible for new AAAS initiatives. Whether as a member of the Board, or as president or chairman, he was a friendly and available uncle to whom a young person could take his or her troubles and come away with a sense of having talked with a friend rather than an authority figure. He was invariably a good listener with the gift for appearing to think his way to answers with the staff person.

When Fred's term ended as an elected officer, we were temporarily stumped about presenting him with a suitable memento. It was Virginia who revealed Fred's secret passion for geometric art. The final dinner that year was held in the great hall of the National Academy of Sciences. It was the same night that Frank Press's wife, Billie, along with Joseph Charyk, led an impromptu audience rendition of "Some Enchanted Evening" in tribute to the memory of Philip Handler, who had died a few days earlier, and Fred's voice was raised with the others while tears brimmed in his eyes. Then we bade him farewell with an ovation and a matched set of geometric prints.

Appendix 2: Resolution adopted by the Board of Trustees of the Russell Sage Foundation

Whereas
Frederick Mosteller

has served as a Trustee of Russell Sage Foundation for twenty-one years during which he has regularly enriched our meetings with clarity of thought; and

Whereas, five Russell Sage presidents and one interrex have been well advised when they have heeded Mosteller; and

Whereas Fred Mosteller, with his unique combination of statistical logic and common sense, has improved our understanding and judgment of the issues that come before us; and

Whereas Mosteller aphorisms, timely jibes, and homilies have become a part of RSF lore; and

Whereas, our friend Fred has not only sharpened our sense of the criteria for rigorous social research, he has also reminded us that risk taking is an inherent part of our enterprise; and

Whereas, the gentleness and sense of wonder with which Fred has listened to the perorations of his fellow trustees has made even the least of us feel worthy;

Resolved therefore, that we, the Trustees of the Russell Sage Foundation, hereby express our profound appreciation to Frederick Mosteller for his years of service in which the Foundation has enjoyed the warmth and candor of a beloved colleague. He has won the respect, admiration and gratitude of his fellow Trustees who now extend to him and to Virginia their personal wishes for a happy future including, we presume, continued intellectual adventures and moments of fun.

November 4, 1985 John S. Reed
New York City Chair

Chapter 4

CONTRIBUTIONS TO MATHEMATICAL STATISTICS

Persi Diaconis and **Erich Lehmann**

At Harvard professors get to choose their titles. Bill Cochran was Professor of Statistics; Art Dempster is Professor of Theoretical Statistics. Since the start of Harvard's department, Frederick Mosteller has been Professor of Mathematical Statistics.

Fred has always been extremely enthusiastic about the place of mathematics in statistics. Time after time he has encouraged us when we saw an obviously mathematical biway we wanted to pursue. At the same time, his success in finding problems where a bit of mathematics made a real-world difference has inspired many other statisticians to seek such problems in their own work.

We have chosen to review four areas of Fred's work. The first two, systematic statistics and slippage tests, border on nonparametrics and robustness. They began with his thesis work and are alive and active today. The third, mathematical learning theory, is a house of which Fred was one of the chief architects. The last, number theory, we include because Fred obviously enjoys it so much.

Our review of his work uncovered some patterns: Fred often is the first to translate an important real-world problem into statistical language. He does significant work on the leading special cases—usually something that both lives on, and convinces the rest of the community that there is a rich vein to be mined.

Our review omits Fred's pioneering work on the secretary problem (P64), which Ferguson (1989) covers nicely. We also do not discuss a host of other fascinating special topics (P3, P7, P48, P62, P70, P77, P87, P140, P143). These papers have also made an impact, stimulating further research. We hope that readers will be tempted to browse in them.

Fred's work on paired comparisons (P18, P19, P20) and his work on contingency tables (B24, P73, B39) are treated in Chapter 5.

4.1 Systematic Statistics

The paper "On some useful 'inefficient' statistics" (P4)—based on Fred's doctoral thesis, written under the supervision of S.S. Wilks—introduces the estimators of location and scale known as *systematic statistics*: finite linear combinations of order statistics

$$L = \sum_{i=1}^{k} w_i X_{(r_i)} \tag{4.1}$$

where $X_{(1)} < \ldots < X_{(n)}$ denotes the ordered sample and where k is fixed, independent of n. Mosteller restricts attention to the case $r_i/n \to \lambda_i$ with $0 < \lambda_i < 1$; i.e., to linear combinations of central order statistics. As a basic tool, he obtains the multivariate normal limiting joint distribution of $X_{(r_1)}, \ldots, X_{(r_k)}$ as $n \to \infty$. His proof is still a standard reference (see David (1981, p. 257)). This limit theorem yields the asymptotic variance of L, which is required for efficiency comparisons.

As a first illustration the paper considers using $w_i = 1/k$ to obtain estimators of the mean of a normal distribution based on only a few order statistics. This section also addresses the problem of optimal spacing; i.e., of finding—for fixed k—the values $r_1, \ldots, r_k$ which minimize the asymptotic variance of L. A second application treats the estimation of the normal standard deviation σ and, in particular, proposes *quasi-ranges*—estimators of the form $c[X_{(n-r+1)} - X_{(r)}]$—for this purpose. The last part of the paper deals with estimation of a normal correlation coefficient from certain quadrant counts.

Mosteller's principal motivation for considering systematic statistics is ease of computation. Robustness against non-normality is mentioned only once, in a comparison with estimators based on extreme observations. In the later literature dealing with these estimators, the emphasis gradually shifts. While ease of computation continues to be stressed, robustness aspects now tend to receive more attention. Estimators from the class proposed by Mosteller are discussed by Benson (1949), Ogawa (1951), Cadwell (1953), Dixon (1957), Harter (1959), Govindarajulu (1963), Bloch (1966), Gastwirth (1966), Crow and Siddiqui (1967), Dubey (1967), Andrews et al. (1972), Chan and Chan (1973), Balmer, Boulton and Sack (1974), Chan and Rhodin (1980), Brown (1981), Eubank (1981), and Lee, Kapadia and Hutcherson (1982).

Evaluation of the exact variance of L (for small samples when the asymptotic theory is not applicable) requires the variances and covariances of the order statistics $X_{(1)}, \ldots, X_{(n)}$. A first effort in this direction was made by Hastings, Mosteller, Tukey, and Winsor in P5, where "the means, variances, and covariances for samples of size ≤ 10 from the normal distribution are tabled and compared with the usual asymptotic approximations." These authors go on in their introduction to state the following program. "It would be very helpful to have (1) at least the first two moments (including product moments) of the order statistics, and (2) tables of the percentage points of their distributions, for samples of sizes from 1 to some moderately large value such as 100 and for a large representative family of distributions. This is a large order and will require much computation." In the intervening years much of this program has been carried out. Summaries (with references) of the enormous literature stemming from both P4 and P5 appear in the books by Sarhan and Greenberg (1962) and David (1981).

4.2 Slippage Tests: "The Problem of the Greatest One"

The classical procedure for comparing k populations tests the hypothesis H of no differences. The decision to accept or reject H is usually inadequate because one needs to go beyond the mere assertion of equality or inequality. Mosteller was the first to suggest that one might want to try to determine the best population. His paper "A k-sample slippage test for an extreme population" (P6) considers samples from k distributions of common shape but with unknown locations $a_1, \ldots, a_k$ and poses the problem of following a test of $H : a_1 = \ldots = a_k$ by the determination of the largest of the a's in case H is rejected. He assumes a common sample size n, and provides the following nonparametric solution to this problem. Let i be the population giving rise to the largest observation, and let the test statistic be the number r of observations in the ith sample which exceed all the observations in the other $k-1$ samples. If $r < r_0$, the null hypothesis of homogeneity is accepted; otherwise, a_i is declared to be the largest of the a's.

In discussing the properties of this procedure, Mosteller considers the following three types of error:

Type I. Rejecting the null hypothesis when it is true. To control the probability of an error of this first kind, he obtains the (exact) null distribution of r and an approximation for large n. He also provides a table of this distribution.

Type II. Accepting the null hypothesis when it is false. Related to this is the power of the procedure, defined as "the probability of both correct rejection and correct choice of [the] rightmost population when it exists." He obtains bounds for the power by considering two cases:

i. $k-1$ of the populations are identical with density $f(x)$, while the remaining population has density $f(x-a)$ with $a > 0$. This provides a lower bound for the power for all cases in which $a_{(k)} - a_{(k-1)} = a$ (where $a_{(1)} \leq \ldots \leq a_{(k)}$ denote the ordered a's).

ii. $a_{(k)} = a, a_{(k-1)} = 0$, and $a_{(1)}, \ldots, a_{(k-2)}$ are so close to $-\infty$

that the associated samples make no contribution to the power. This gives an upper bound for the power for all cases in which $a_{(k)} - a_{(k-1)} = a$.

Two short tables provide some numerical examples of these upper and lower bounds for the case $k = n = 3$.

Type III. Correctly rejecting the null hypothesis for the wrong reasons. This error of the third kind is mentioned but not evaluated.

Mosteller points out that the problem of interest is not really that of testing H, but rather that of determining the population with the largest a-value, dubbed by him "the problem of the greatest one." The null hypothesis H is an artifice (Mosteller calls it a trick) to provide some control in case not much is actually going on.

Mosteller outlines a program of further research, mentioning three specific problems:

A. To obtain a test more powerful than the one proposed in the paper.

B. To consider the ordering of several populations.

C. To formulate and study different alternatives, null hypotheses, and test statistics.

In the last sentence, he expresses the wish "that more material on these problems will appear, because answers to these questions are urgently needed in practical problems." This wish has been fulfilled abundantly, largely in response to the stimulus provided by this paper and its sequel, P15, which extends the procedure to unequal sample sizes. In fact, the enormous literature that has grown up around this program contains practically no paper that cannot be connected with P6 through a chain of references.

As item A indicates, Mosteller himself felt that his procedure could probably be improved upon. Paulson (1952b) considered the problem of determining an optimum procedure. To this end, he made two specializations in Mosteller's formulation. (i) He restricted attention to a parametric model: k samples of equal size n from normal distributions with common variance. (ii) He maximized the power against the alternative that one of the k populations "has slipped to the right by an amount Δ." Under these assumptions, he was able to show that there exists a uniformly most powerful invariant (under a suitable group of transformations) procedure based on the statistic $(\bar{x}_M - \bar{x})/S$, where $\bar{x}_M$ is the largest of the k sample means, $\bar{x} = \Sigma\Sigma x_{ij}/kn$ and $S^2 = \Sigma\Sigma(x_{ij} - \bar{x})^2/kn$.

The corresponding problem for normal variances was treated by Truax (1953), and further applications and various generalizations of this "slippage problem" were discussed by Paulson (1952a), Kudo (1956), Doornbos and Prins (1956, 1958), Kapur (1957), Ramachandran and Khatri

(1957), Pfanzagl (1959), Karlin and Truax (1960), Paulson (1961), Roberts (1964), Bofinger (1965), Doornbos (1966), Conover (1968), Hall and Kudo (1968), Hall, Kudo and Yeh (1968), Kakiuchi and Kimura (1957), Kakiuchi, Kimura, and Yanagawa (1977), Hashemi-Parast and Young (1979), Joshi and Sathe (1981), and Kimura (1984). Sequential procedures were proposed, for example, by Paulson (1962), Roberts (1963), and Srivastava (1973). An excellent review of this body of work is given by Barnett and Lewis (1984, Chapter 8); see also Schwager (1985).

The formulation used by most of these authors suffers in realism because it restricts attention to the situation in which it is known that only one population has slipped.[1] Additional problems arise from the somewhat dubious role played by the null hypothesis. These difficulties can be avoided by reverting to Mosteller's original problem of "the greatest one"—i.e., determination of the "best" population, for example, that having the largest a-value. The following are a few of the early papers dealing with this problem directly (i.e., without reference to any null hypothesis) by specifying the loss resulting from the selection of a non-optimal population, and then using invariance, Bayes, or minimax approaches to obtain an optimal decision rule: Bahadur (1950), Bechhofer (1954), Somerville (1954, 1970), Grundy, Healy and Rees (1956), Dunnett (1960), and Guttman and Tiao (1964). The two principal influences which stimulated the initial work on this problem are Mosteller's paper (P6) and the general framework provided by Wald's decision theory.

A natural generalization of the problem of "the greatest one" initiated by Bechhofer (1954) is that of specifying the t best populations (either ordered or unordered), which for $t = k$ becomes the problem of ranking k populations. This in turn can be viewed as a particular goal of multiple comparisons; the connection is explored, for example, in Neave (1975). (See also Miller (1981, Sec. 6.5).)

As an alternative to controlling error via a null hypothesis, as in Mosteller's original formulation, Gupta (1956, 1965) proposed to select a group of populations, whose size is determined by the observations and which contains the best population with probability $\geq \gamma$ (specified). An enormous literature on such selection procedures (one-stage, two-stage, and fully sequential; parametric and nonparametric) has developed, and the subject—together with that of ranking a number of populations—is treated in full detail in the books by Gibbons, Olkin, and Sobel (1977), *Selecting and Ordering Populations: A New Statistical Methodology*, and by Gupta and Panchapakesan (1979), *Multiple Decision Procedures: Theory*

[1]Actually, in some of the cases this assumption can be avoided by adopting the maximin approach of maximizing the minimum power subject to $a_{(k)} - a_{(k-1)} \geq \Delta$. It then frequently turns out that the least favorable configurations are characterized by $a_{(1)} = \ldots = a_{(k-1)} = a_{(k)} - \Delta$, as Mosteller pointed out for his procedure in II(i).

and Methodology of Selecting and Ranking Populations. Gupta and Huang (1981) survey more recent developments.

Mosteller's 1948 paper has thus sparked three subjects, each with a large literature: Slippage tests, ranking of populations, and selection. A fourth subject, although closely related, has developed quite separately: the detection of outlying observations. In its simplest form, the problem here concerns a sample $X_1, \ldots, X_k$ and the question of whether, say, an abnormally large observation comes from the same distribution as the rest of the observations. This is just the slippage problem with sample size $n = 1$ (in which, however, no estimate of variance or other measure of internal variability is available). Conversely, if a slippage problem is studied in terms of statistics $T_i = T_i(X_{i1}, \ldots, X_{in})$, it becomes equivalent to an investigation of outliers in the observations $T_1, \ldots, T_k$. It is thus not surprising that in the literature on the outlier problem, which goes back to the 19th century and is reviewed in the books by Barnett and Lewis (1984) and by Hawkins (1980), one finds many of the ideas that occur in the slippage and selection literature. What is surprising is how little these two literatures have influenced each other.[2] The explanation is presumably the difference in context and purpose, indicated by the question: "Is this observation too large to fit in with the rest of the sample?" on the one hand, and Mosteller's problem of determining the treatment with the largest effect on the other.

4.3 Stochastic Models for Learning

How rapidly does learning occur under various circumstances? This is a basic question of psychology. Fred Mosteller, working with Robert Bush, spearheaded a new analytic approach to these problems called mathematical learning theory which held center stage in this area for a fifteen-year period. Their work was followed by hundreds of further studies and has led to some lasting insights. It may be helpful to review some aspects of learning theory before describing Mosteller's work.

A Bit of Learning Theory

Imagine facing a box with two lights and two buttons.

You push either the left or right button. Then one of the two lights flashes on. If your guess is correct, you get a dollar; otherwise you get nothing.

In one widely replicated version of this two-choice experiment the box is programmed to simulate coin tossing with probability .7. Thus the left light turns on at random 70% of the time; the right light turns on the other 30%. And there is no other pattern. It is natural to inquire how long it will take

[2] Among the few papers considering both are Ferguson (1961) and Butler (1981).

a subject to learn this and adopt the optimal strategy—always push the left button. Some of the major psychological findings from this two-choice experiment are summarized as follows:

- Most subjects never adopt the optimal strategy, even after hundreads of trials.

- Typically, subjects roughly conform to "probability matching": they wind up guessing left 70% of the time (or $100p$ percent if p is the observed proportion). Moreover, if they are asked to estimate directly the probability of the left light, their estimates approach the true frequency.

- Subjects invent strategies based on "patterns" in the data such as "a right after two rights and a left except after a run of 3."

- Less intelligent guessers (pigeons, goldfish) do better. Really dumb organisms eventually learn always to choose the higher-payoff alternative.

- Increasing the monetary reward helps, perhaps changing the subject's proportion of left guesses from 70% to 80%. Most subjects never get up to 100%.

Of course, simplifications such as these must be carefully quantified. An authoritative recent account of modern learning theory is in Bower and Hilgard (1981). Chapter 13 of Mosteller's book with Bush (B5) and the paper of Norman and Yellott (1966) contain detailed discussion of two-choice experiments.

The list above contains some surprises. The first two items show that we don't always behave optimally in making decisions under uncertainty, and there may be patterns in the way we do behave. Kahneman, Slovic, and Tversky (1982) is a good place to find more on this subject. The next item suggests that our tendency to see patterns in noise must be taken into account in evaluating routine data analysis. (See Chapter 1 of B54 for more.) The final item suggests that subjects in psychological experiments can respond differently as motivation changes.

The two-choice experiment yields binary data in the subjects' pattern of responses. The type of binary data making up the pattern of responses in a two-choice experiment occurs also in other learning experiments. In a T-maze experiment a rat runs down a ramp and chooses the left fork or right fork, receiving food some of the time. In verbal learning experiments, a list of words is read to a subject (the order being scrambled from trial to trial). After this, the subjects write down all remembered words. Following a particular word from trial to trial yields binary outcomes as the word is recalled or not.

When Mosteller began work in this area, there was a sea of data but little quantitative work. One of his main contributions centers around a basic class of models, described next.

The Bush-Mosteller Model

The original Bush-Mosteller paper (P25) begins as follows:

> Mathematical models for empirical phenomena aid the development of a science when a sufficient body of quantitative information has been accumulated. This accumulation can be used to point the direction in which models should be constructed and to test the adequacy of such models in their interim states. Models, in turn, frequently are useful in organizing and interpreting experimental data and in suggesting new directions for experimental research. Among the branches of psychology, few are as rich as learning in quantity and variety of available data necessary for model building.

The paper goes on to describe a model for the two-response experiments described above. A basic construct of the model is a parameter p_n—the probability of guessing "left" on the nth trial. When the guess is correct, the increase from p_n to p_{n+1} could be modeled via

$$p_{n+1} = p_n + \gamma_n \tag{4.2}$$

where γ_n is a positive multiple of the largest possible change that stays between p_n and 1, so that $\gamma_n = a(1 - p_n)$ for fixed a, $0 < a < 1$. Thus the change is modeled as the affine function

$$p_{n+1} = (1 - a)p_n + a. \tag{4.3}$$

Generalizing slightly, the Bush-Mosteller model considers affine changes of the form

$$p_{n+1} = \alpha p_n + \beta \quad \text{where} \quad 0 < \beta < 1, -\beta < \alpha < 1 - \beta. \tag{4.4}$$

It is natural to let α and β depend on whether a left or right light occurred on trial n:

$$p_{n+1} = \begin{cases} \alpha_1 p_n + \beta_1 & \text{when no error is made on trial } n \\ \alpha_2 p_n + \beta_2 & \text{when an error is made on trial } n. \end{cases} \tag{4.5}$$

This, together with the starting value p_0, specifies the basic five-parameter Bush-Mosteller model.

The first paper motivated the model above as a linear approximation to a more general change. Much of the paper deals with the important special case where the left light always goes on. The discrete model and associated difference equation are approximated by a continuous model and more manageable differential equations. The model is shown to give a reasonable fit to a real data set.

Throughout, the paper makes careful comparison with a model introduced by the psychologist William Estes (1950). This has $\beta_2 = 0$, $\alpha_1 = \alpha_2 = \alpha$, and $\beta_1 = 1 - \alpha$. Estes begins from a different set of primitive assumptions. Bush and Mosteller derive their affine models from Estes's assumption in Chapter 5 of their book (B5).

The second paper (P30) provides a more theoretical exploration of the five-parameter model. In mathematical language, the sequence $p_0, p_1, p_2, \ldots$ evolves as a stochastic process, varying according the sequence of left and right lights observed. Of course, the sequence $\{p_n\}$ is not observed, merely the guess on each trial. This raises a host of questions: How will the p_n and observed guessing process evolve? How should the parameters $p_0, \alpha_1, \beta_1, \alpha_2, \beta_2$ be estimated from observed guesses? How can the validity of the model be checked?

Significant progress on these questions is reported in P30. Under some assumptions, the model predicts that, in a two-choice experiment with the left light on proportion p of the time, the proportion of subjects' left guesses tends to p. Thus, the model offers an explanation for the strategy of "probability matching." The behavior of the p_n process is also studied by getting bounds and some closed-form expressions for moments. Although no general solution for the estimation problem is presented, two important special cases are identified:

- the two affine operators in the definition of p_{n+1} above commute
- the two slopes are equal ($\alpha_1 = \alpha_2$)

The method of maximum likelihood is applied in these cases. The paper explains how the model and analysis apply to a wide variety of binary learning experiments.

A later paper (P45) studies the behavior of the p_n process for a commuting model:

$$p_{n+1} = \begin{cases} \alpha_1 p_n & \text{when no error is made on trial } n \\ \alpha_2 p_n & \text{when an error is made on trial } n. \end{cases} \tag{4.6}$$

Then the probability of error tends to be zero. It is shown that only finitely many errors occur, and the distribution of the last error is discussed.

In their book, Bush and Mosteller provided a systematic account of what was then known about mathematical learning theory. The book offers a clear tutorial on the tools of probability and statistics applied to the problems of learning theory. The model and recent extensions are developed in detail. Properties of the model are checked by Monte Carlo experiments, and by application to real data. The model is used as a base for insightful data analysis of several experiments that had previously appeared in the psychology literature. New data, collected by the authors, are also analyzed. Further discussion of the book appears in Paul Holland's review in Chapter 8 of the present volume.

The book is a serious intellectual effort by two eminent scholars. It establishes a useful connection between psychology and modern mathematics (differential and difference equations). The authoritative applications of probability and statistics to psychological phenomena initiated a new approach for the field.

The book's clarity, together with the complexity of the basic learning experiment, brought on a torrent of further work, competing models, and ever more complex data. The next stage of the story is best considered in a historical setting.

Some History

The quantitative study of learning has roots in the 19th century. See, for example, Miller (1964) or Atkinson, Bower, and Crothers (1965, Sec. 1.5). Mosteller's work on the subject builds on efforts of psychologists such as Thurstone (1919) and Hull (1952). Thurstone considered an unobservable p_n process, and Hull attempted to derive learning curves from basic postulates. Neither of these early attempts was crisply quantified.

Mosteller began work on the model in 1949, when psychology was ruled by behaviorism—the study of performance and its change through direct observation. This was also a time of great optimism for the use of mathematical models in the social sciences. The Social Science Research Council was actively involved in funding such projects, and in encouraging the "retraining" of physicists like Robert Bush. Game theory was developing rapidly. Mosteller (P95, P97) records reminiscences of these early years.

Mosteller worked in parallel with eminent psychologists such as William Estes, George Miller, and Duncan Luce. They had as a common goal the building of an intellectual foundation for investigations of learning. Arriving at similar models from different bases, they made and explored connections in joint work, at conferences, and in multiple-authored volumes.

Scores of other researchers joined the bandwagon, making learning theory of the 1950s and 1960s a very active research area. Experimental work was carried out along with theory, summarized in books by Atkinson, Bower, and Crothers (1965), Iosifescu and Theodorescu (1969), and Norman (1972), as well as in dozens of conference proceedings and hundreds of papers.

Then, almost as suddenly as it began, work in mathematical learning slowed to a trickle. This occurred so abruptly that young mathematical psychologists who emerged at that time report feelings of confusion and shock.

Many forces contributed to the end. Within the field itself, although a number of different models could successfully predict the proportion of correct guesses over time averaged over subjects, none had great success at matching a single subject's sequential behavior. Upon closer inspection, details of the model's account of probability-learning turned out to be wrong. Subjects were trying out a variety of complex sequential hypotheses about the left-right light series; although such hypotheses resulted in probability matching, it was for reasons unrelated to those suggested by the model.

Many of the most striking psychological findings from two-choice experiments (listed above) had been empirically discovered before mathematical modeling. The models did not add many new phenomena.

The interplay between mathematics and experiment forced psychologists to recognize that overall learning was far more complex than the sum of many simple processes. Simple experiments seemed to give no real clue to understanding human learning in the large. Today, it still seems impossible to give reasonable models (or even state-space descriptions) for learning. We don't know enough about which variables are basic.

An important force from outside learning theory that contributed to the end was the sweeping shift from operationalism to cognitive psychology. In the operational paradigm, talk of "the mind" was banned. In the cognitive paradigm it is perfectly all right to speculate about the mind, as long as such speculations are matched with predictions and experiments.

The cognitive paradigm brought new kinds of questions, a shift in interest from learning to memory, and a distrust of old tools such as mathematical modeling. Game theory stopped dead in its tracks about the same time as mathematical learning theory. Without the support of psychologists, other theorists withdrew, and work in the field was discontinued.

Afterward

The work initiated by Mosteller continues to have an impact in psychology. To begin with, it helped codify the basic psychological findings listed above. Estes's "stimulus sampling theory" is a standard part of current learning theory. See, for example, Bower and Hilgard (1981, Chapter 8).

The history sketched above helps us understand the limitations of statistical modeling when the underlying "physics" of a field has not been sorted out. Even when practiced in the most serious fashion, statistical modeling should not be expected to perform miracles.

Mosteller's work on learning theory is the first clear example of modern mathematical psychology. It stimulated work in several other areas of psychology, as summarized by Miller (1964). Some of these, most notably choice theory, are very much alive and healthy today. In addition to this, Mosteller's work leaves a rich mathematical legacy which has taken on a life of its own. Three examples are given below.

4.4 Products of Random Matrices, Computer Image Generation, and Computer Learning

The Bush-Mosteller model has a close connection with recent work on products of random matrices. To see this, write the basic affine transformation $p_{n+1} = \alpha p_n + \beta$ in matrix form

$$\begin{pmatrix} \alpha & \beta \\ 0 & 1 \end{pmatrix} \begin{pmatrix} p_n \\ 1 \end{pmatrix} = \begin{pmatrix} \alpha p_n + \beta \\ 1 \end{pmatrix} = \begin{pmatrix} p_{n+1} \\ 1 \end{pmatrix} \tag{4.7}$$

Suppose the nth pair (α^n, β^n) is chosen as (α_0, β_0) with probability p and (α_1, β_1) with probability $1-p$, independently at each stage. Then

$$\begin{pmatrix} p_{n+1} \\ 1 \end{pmatrix} = \begin{pmatrix} \alpha^n & \beta^n \\ 0 & 1 \end{pmatrix} \begin{pmatrix} \alpha^{n-1} & \beta^{n-1} \\ 0 & 1 \end{pmatrix} \cdots \begin{pmatrix} \alpha^1 & \beta^1 \\ 0 & 1 \end{pmatrix} \begin{pmatrix} p_0 \\ 1 \end{pmatrix} \tag{4.8}$$

Thus, methods of studying products of random matrices, or more generally, random walks on groups, can be used to analyze the behavior of p_n.

Historically, it worked the other way. Methods developed by Norman (1972) and Kaijser (1978) to analyze the Bush-Mosteller model and its generalizations have proved central in the recent work of the French school analyzing random walks on groups. This work is elegantly presented in the recent monograph by Bougerol and Lacroix (1985), who make the relevant connections.

The p_n process described above evolves as a Markov chain. Bush, Mosteller, and colleagues such as Karlin (1953), Garsia (1962), Dubins and Freedman (1966), and many others studied this Markov chain quite carefully. The same chain has recently been applied to a simple algorithm for generating complex pictures of leaves, trees, clouds, and other natural objects on a computer. The algorithm uses a finite number of affine transformations to generate a Markov chain (say, in the plane). The path traced out by the chain becomes the picture. Details are contained in Demko et al. (1985), Barnsley et al. (1986), and Diaconis and Shahshahani (1986).

Recently, fresh interest in mathematical learning theory has been generated by the efforts of the computer science community to program computers to learn. Computer learning often proceeds by simple reinforcement, so that assumptions of models like Bush-Mosteller are exactly correct. Classical studies of properties of the model can now be interpreted as rates of convergence. Surveys of this literature appear in Lakshmivarahan (1981) and Herkenrath, Kalin, and Vogel (1983).

For the present discussion, the point is this: Mosteller initiated serious work on affine models. This work led to much research, which is being usefully applied, 25 years later, in completely new areas.

4.5 Number Theory—Statistics for the Love of It

Over the years, Mosteller has had a love affair with the prime numbers. He studies them as data, finding patterns and making probabilistic and mathematical sense of the patterns. He often presents this work as material for teaching. In P90 he wrote

> It occurred to me that students had available an unlimited supply of statistics, though not always cheaply, in the properties of the integers. For example, the distribution of the early primes (Mosteller, 1972) (P91), the frequency distribution of the numbers of prime divisors (Mosteller, 1971) for the first n integers. A fair amount of statistics can be learned by trying to bring some order into the facts about these problems even for the first few (hundreds of) integers.

His studies have produced both interesting teaching material and interesting mathematics. In some instances, Mosteller has guessed at provable patterns, leading to new number theory. Some of this is discussed further below. At least as important as these results is the joy that Mosteller finds and communicates in discussing number theory. Both the number theory and the statistics come alive to a wide audience.

Primes as Data

The prime numbers, $2, 3, 5, 7, 11, \ldots$, form a notoriously irregular sequence. There is no "formula" for predicting the nth prime. The best we can do is to demonstrate statistical regularities. For example, the prime number theorem asserts that the proportion of integers between 1 and n that are prime is approximately $1/\log_e n$.

If this were the only regularity, then primes in intervals should be randomly distributed, like balls dropped into boxes or points from a Poisson process. To study this assumption, Mosteller divided the numbers from 10×10^6 through 12.5×10^6 into 50 intervals of length 50,000, and for each of the 50 intervals, computed the number of primes in the interval. These 50 numbers have mean 3082.48 and variance 615.20. If the primes showed Poisson variation, the mean would be close to the variance. Such relatively small variances occur most of the time, suggesting some further structure. Any regularity in the primes is surprising. Mosteller explored and quantified things in a fascinating data analysis.

The structure seems to be related to twin primes, a second interest of Mosteller. The twin prime conjecture states that there are infinitely many primes p such that $p + 2$ is also prime. Similar conjectures cover prime triples and prime r-tuples. If the primes satisfy the additional constraints of the prime r-tuples conjecture, there is then enough negative correlation to cause the variance to decrease. Gallagher (1976) assumed the truth of the prime r-tuples conjecture and derived the moments of the number of primes in intervals. Since Gallagher worked with relatively short intervals, the structure observed by Mosteller is not evident.

Often hard problems about the primes can be translated into easier problems about square-free numbers. An integer n is square-free if it has no square prime divisors, so 2, 3, 5, 6, 7, 10, 11, 13 are square-free, whereas 4, 8, 9, 12 are not. The proportion of square-free numbers is $6/\pi^2$ in the limit. For square-free numbers, it can be proved that the structure observed by Mosteller is really there.

Charles Stein (1981) and R.R. Hall (1982) proved that the variance of the square-free numbers in intervals is too small without assuming any unproved hypothesis. Stein, who acknowledges that Mosteller's data analysis set him off, speculates further: perhaps the number of square-free numbers in consecutive intervals can be normed to converge to a fractional Brownian motion with parameter 1/2. This iteration between data analysis and proof makes it hard to answer the question "Is mathematics a science?"

The Number of Prime Divisors

One property of an integer n is its number of distinct prime divisors $\omega(n)$. For example, 12 has divisors 2 and 3, so $\omega(12) = 2$. Empirical investigation reveals that most numbers we encounter have very few prime divisors. It is natural to ask questions about the distribution of prime divisors. Hardy and Littlewood obtained the first results in the 1920's. They showed that most numbers n had

$$\omega(n) \doteq \log_e \log_e n \quad . \tag{4.9}$$

Now $\log\log 10^6 \doteq 2.63$, $\log\log 10^9 \doteq 3.03$, so most numbers we routinely encounter have fewer than 3 distinct prime divisors.

Erdös and Kac proved that the distribution is approximately normal: the proportion of $n \leq N$ with

$$\frac{\omega(n) - \log\log n}{\sqrt{\log\log n}} \leq t \tag{4.10}$$

tends to the area under the normal curve to the left of t. Their paper is also a landmark of modern probability as the first appearance of what is now called the invariance principle. Kac (1959), Elliott (1979, 1980), and Billingsley (1974) review this work.

Mosteller (1971) investigated the distribution of prime divisors empirically, showing how a Poisson approximation leapt out if one looked at things in the right way. He also examined the accuracy of the asymptotic formulas of Hardy and Ramanujan. The formulas didn't seem so useful for n as small as 10^9. Indeed, when Diaconis first visited Harvard in deciding about graduate school, Mosteller looked over his transcript, saw an interest in number theory, and asked for help in deriving a correction term for the variance of $\omega(n)$. This led to joint work of Diaconis, Mosteller, and Onishi (P104). Diaconis (1976) continued work on the problem, which also led to the Ph.D. thesis of Rejali (1978).

The methods derived to understand Mosteller's original question even had some practical impact. Diaconis (1980) used them to derive the average running time of the fast Fourier transform. The Cooley-Tukey version of this algorithm uses the prime factorization of the series length n and runs in a number of operations roughly equal to n times the sum of the primes that divide n. The number of operations, as a function of n, is sufficiently similar to $\omega(n)$ that old techniques gave new results. Computations of average running time suggest that the Cooley-Tukey algorithm is not as good as others (such as the chirp-z transform).

This gives a nice example of mathematics, done for the love of it, yielding insight into a real-world problem.

References

Andrews, D.F., Bickel, P.J., Hampel, F.R., Huber, P.J., Rogers, W.H., and Tukey, J.W. (1972). *Robust Estimates of Location.* Princeton, NJ: Princeton University Press.

Atkinson, R.C., Bower, G.H., and Crothers, E.J. (1965). *An Introduction to Mathematical Learning Theory.* New York: Wiley.

Bahadur, R.R. (1950). On a problem in the theory of k populations. *Annals of Mathematical Statistics*, **21**, 362–375.

Balmer, D.W., Boulton, M., and Sack, R.A. (1974). Optimal solutions in parameter estimation problems for the Cauchy distribution. *Journal of the American Statistical Association*, **69**, 238–242.

Barnett, V. and Lewis, T. (1984). *Outliers in Statistical Data.* 2nd Ed. New York: Wiley.

Barnsley, M.F., Ervin, V., Hardin, D.P., and Lancaster, J. (1986). Solution of an inverse problem for fractals and other sets. *Proceedings of the National Academy of Sciences*, **83**, 1975–1977.

Bechhofer, R.E. (1954). A single-sample multiple decision procedure for ranking means of normal populations with known variances. *Annals of Mathematical Statistics*, **25**, 16–39.

Benson, F. (1949). A note on the estimation of mean and standard deviation from quantiles. *Journal of the Royal Statistical Society, Series B*, **11**, 91–100.

Billingsley, P. (1974). The probability theory of additive arithmetic functions. *Annals of Probability*, **2**, 749–791.

Bloch, D. (1966). A note on the estimation of the location parameter of the Cauchy distribution. *Journal of the American Statistical Association*, **61**, 852–855.

Bofinger, V.J. (1965). The k-sample slippage problem. *Australian Journal of Statistics*, **7**, 20–31.

Bougerol, P. and Lacroix, J. (1985). *Products of Random Matrices with Applications to Schrödinger Operators.* Boston: Birkhäuser.

Bower, G.H. and Hilgard, E.R. (1981). *Theories of Learning.* 5th Ed. Englewood Cliffs, NJ: Prentice-Hall.

Brown, B.M. (1981). Symmetric quantile averages and related estimators. *Biometrika*, **68**, 235–242.

Butler, R.W. (1981). The admissible Bayes character of subset selection techniques involved in variable selection, outlier detection, and slippage problems. *Annals of Statistics*, **9**, 960–973.

Cadwell, J.H. (1953). The distribution of quasi-ranges in samples from a normal population. *Annals of Mathematical Statistics*, **24**, 603–613.

Chan, L.K. and Chan, N.N. (1973). On the optimum best linear unbiased estimates of the parameters of the normal distribution based on selected order statistics. *Skandinavisk Aktuarietidskrift*, **1973**, 120–128.

Chan, L.K. and Rhodin, L. (1980). Robust estimation of location using optimally chosen sample quantiles. *Technometrics*, **22**, 225–237.

Conover, W.J. (1968). Two k-sample slippage tests. *Journal of the American Statistical Association*, **63**, 614–626.

Crow, E.L. and Siddiqui, M.M. (1967). Robust estimation of location. *Journal of the American Statistical Association*, **62**, 353–389.

David, H.A. (1981). *Order Statistics.* 2nd Ed. New York: Wiley.

Demko, S., Hodges, L., and Naylor, B. (1985). Construction of fractal objects with iterated function systems. *Computer Graphics*, **19**, No. 3, 271–278. (*SIGGRAPH '85 Proceedings*).

Diaconis, P. (1976). Asymptotic expansions for the mean and variance of the number of prime factors of a number n. Technical Report No. 96, Department of Statistics, Stanford University.

Diaconis, P. (1980). Average running time of the fast Fourier transform. *Journal of Algorithms*, **1**, 187–208.

Diaconis, P. and Shahshahani, M. Products of random matrices and computer image generation. In *Random Matrices and Their Applications* (Contemporary Mathematics, Vol. 50), edited by J.E. Cohen, H. Kesten, and C.M. Newman. Providence, RI: American Mathematical Society, 1986. pp. 173–182.

Dixon, W.J. (1957). Estimates of the mean and standard deviation of a normal population. *Annals of Mathematical Statistics*, **28**, 806–809.

Doornbos, R. (1966). Slippage tests. Mathematical Centre Tracts, No. 15, Mathematisch Centrum, Amsterdam.

Doornbos, R. and Prins, H.J. (1956). Slippage tests for a set of gamma-variates. *Indagationes Mathematicae*, **18**, 329–337.

Doornbos, R. and Prins, H.J. (1958). On slippage tests. *Indagationes Mathematicae*, **20**, 38–55, 438–447.

Dubey, S.D. (1967). Some percentile estimators for Weibull parameters. *Technometrics*, **9**, 119-129.

Dubins, L.E. and Freedman, D. (1966). Invariant probabilities for certain Markov processes. *Annals of Mathematical Statistics*, **37**, 837–848.

Dunnett, C.W. (1960). On selecting the largest of k normal population means (with discussion). *Journal of the Royal Statistical Society, Series B*, **22**, 1–40.

Elliott, P.D.T.A. (1979). *Probabilistic Number Theory I: Mean-Value Theorems.* New York: Springer-Verlag.

Elliott, P.D.T.A. (1980). *Probabilistic Number Theory II: Central Limit Theorems.* New York: Springer-Verlag.

Estes, W. (1950). Toward a statistical theory of learning. *Psychological Review*, **57**, 94–107.

Eubank, R.L. (1981). Estimation of the parameters and quantiles of the logistic distribution by linear functions of sample quantiles. *Scandinavian Actuarial Journal*, **1981**, 229–236.

Ferguson, T.S. (1961). On the rejection of outliers. In *Proceedings of the Fourth Berkeley Symposium on Mathematical Statistics and Probability*, **Vol. 1**, edited by J. Neyman. Berkeley and Los Angeles: University of California Press, 1961. pp. 253–287.

Ferguson, T.S. (1989). Who solved the secretary problem? *Statistical Science*, **4**, 282–296.

Gallagher, P.X. (1976). On the distribution of primes in short intervals. *Mathematika*, **23**, 4–9.

Garsia, A. (1962). Arithmetic properties of Bernoulli convolutions. *Transactions of the American Mathematical Society*, **102**, 469–482.

Gastwirth, J.L. (1966). On robust procedures. *Journal of the American Statistical Association*, **61**, 929–948.

Gibbons, J.D., Olkin, I., and Sobel, M. (1977). *Selecting and Ordering Populations.* New York: Wiley.

Govindarajulu, Z. (1963). On moments of order statistics and quasi-ranges from normal populations. *Annals of Mathematical Statistics*, **34**, 633–651.

Grundy, P.M., Healy, M.J.R., and Rees, D.H. (1956). Economic choice of the amount of experimentation. *Journal of the Royal Statistical Society, Series B*, **18**, 32–49.

Gupta, S.S. (1956). On a decision rule for a problem in ranking means. Ph.D. Thesis (Mimeo. Ser. No. 150). Inst. of Statist., Univ. of North Carolina, Chapel Hill.

Gupta, S.S. (1965). On some multiple decision (selection and ranking) rules. *Technometrics*, **7**, 225–245.

Gupta, S.S. and Huang, D.-Y. (1981). *Multiple Statistical Decision Theory: Recent Developments.* New York: Springer-Verlag.

Gupta, S.S. and Panchapakesan, S. (1979). *Multiple Decision Procedures: Theory and Methodology of Selecting and Ranking Populations.* New York: Wiley.

Guttman, I. and Tiao, G.C. (1964). A Bayesian approach to some best population problems. *Annals of Mathematical Statistics*, **35**, 825–835.

Hall, I.J. and Kudo, A. (1968). On slippage tests. I. *Annals of Mathematical Statistics*, **39**, 1693–1699.

Hall, I.J. Kudo, A., and Yeh, N.C. (1968). On slippage tests. II. *Annals of Mathematical Statistics*, **39**, 2029–2037.

Hall, R.R. (1982). On the distribution of square free numbers in short intervals. *Mathematika*, **29**, 57–68.

Harter, H.L. (1959). The use of sample quasi-ranges in estimating population standard deviation. *Annals of Mathematical Statistics*, **30**, 980–999. Correction: **31**, 228.

Hashemi-Parast, S.M. and Young, D.H. (1979). Distribution free slippage tests following a Lehmann model. *Journal of Statistical Computation and Simulation*, **8**, 237–251.

Hawkins, D.M. (1980). *Identification of Outliers.* London and New York: Chapman and Hall.

Herkenrath, U., Kalin, D., and Vogel, W. (Eds.) (1983). *Mathematical Learning Models—Theory and Algorithms.* New York: Springer-Verlag.

Hull, C.L. (1952). *A Behavior System.* New Haven: Yale University Press.

Iosifescu, M. and Theodorescu, R. (1969). *Random Processes and Learning.* New York: Springer-Verlag.

Joshi, S. and Sathe, Y.S. (1981). A k-sample slippage test for location parameter. *Journal of Statistical Planning and Inference*, **5**, 93–98.

Kac, M. (1959). *Statistical Independence in Probability, Analysis and Number Theory.* (Carus Mathematical Monograph 12, Mathematical Association of America). New York: Wiley.

Kahneman, D., Slovic, P., and Tversky, A. (Eds.) (1982). *Judgment under Uncertainty: Heuristics and Biases.* Cambridge: Cambridge University Press.

Kaijser, T. (1978). A limit theorem for Markov chains in compact metric spaces with application to products of random matrices. *Duke Mathematical Journal*, **45**, 311–349.

Kakiuchi, I. and Kimura, M. (1975). On slippage rank tests. I. *Bulletin of Mathematical Statistics*, **16**, 55–71.

Kakiuchi, I., Kimura, M., and Yanagawa, T. (1977). On slippage rank tests. II. *Bulletin of Mathematical Statistics*, **17**, 1–13.

Kapur, M.N. (1957). A property of the optimum solution suggested by Paulson for the k-sample slippage problem for the normal distribution. *Indian Society of Agricultural Statistics*, **9**, 179–190.

Karlin, S. (1953). Some random walks arising in learning models I. *Pacific Journal of Mathematics*, **3**, 725–756.

Karlin, S. and Truax, D. (1960). Slippage problems. *Annals of Mathematical Statistics*, **31**, 296–324.

Kimura, M. (1984). The asymptotic efficiency of conditional slippage tests for exponential families. *Statistics and Decisions*, **2**, 225–245.

Kudo, A. (1956). On the invariant multiple decision procedures. *Bulletin of Mathematical Statistics*, **6**, 57–68.

Lakshmivarahan, S. (1981). *Learning Algorithms: Theory and Applications.* New York: Springer-Verlag.

Lee, K.R., Kapadia, C.H., and Hutcherson, D. (1982). Statistical properties of quasi-range in small samples from a gamma density. *Communications in Statistics Part B—Simulation and Computation*, **11**, 175–195.

Miller, G.A. (1964). *Mathematics and Psychology.* New York: Wiley.

Miller, R.G., Jr. (1981). *Simultaneous Statistical Inference.* 2nd Ed. New York: Springer-Verlag.

Mosteller, F. (1971). *Fifty Challenging Problems in Probability with Solutions.* Russian translation, pp. 23, 98–101. [See B21 in Bibliography.]

Neave, H.R. (1975). A quick and simple technique for general slippage problems. *Journal of the American Statistical Association*, **70**, 721–726.

Norman, M.F. (1972). *Markov Processes and Learning Models.* New York: Academic Press.

Norman, M.F. and Yellott, J.I., Jr. (1966). Probability matching. *Psychometrika*, **31**, 43–60.

Ogawa, J. (1951). Contributions to the theory of systematic statistics, I. *Osaka Journal of Mathematics*, **3**, 175–213.

Paulson, E. (1952a). On the comparison of several experimental categories with a control. *Annals of Mathematical Statistics*, **23**, 239–246.

Paulson, E. (1952b). An optimum solution to the k-sample slippage problem for the normal distribution. *Annals of Mathematical Statistics*, **23**, 610–616.

Paulson, E. A non-parametric solution for the k-sample slippage problem. In *Studies in Item Analysis and Prediction*, edited by H. Solomon. Stanford, CA: Stanford University Press, 1961. pp. 233–238.

Paulson, E. (1962). A sequential procedure for comparing several experimental categories with a standard or control. *Annals of Mathematical Statistics*, **33**, 438–443.

Pfanzagl, J. (1959). Ein kombiniertes test und klassifikations-problem. *Metrika*, **2**, 11–45.

Ramachandran, K.V. and Khatri, C.G. (1957). On a decision procedure based on the Tukey statistic. *Annals of Mathematical Statistics*, **28**, 802–806.

Rejali, A. (1978). On the asymptotic expansions for the moments and the limiting distributions of some additive arithmetic functions. Ph.D. dissertation, Department of Statistics, Stanford University.

Roberts, C.D. (1963). An asymptotically optimal sequential design for comparing several experimental categories with a control. *Annals of Mathematical Statistics*, **34**, 1486–1493.

Roberts, C.D. (1964). An asymptotically optimal fixed sample size procedure for comparing several experimental categories with a control. *Annals of Mathematical Statistics*, **35**, 1571–1575.

Sarhan, A.E. and Greenberg, B.G. (Eds.). (1962). *Contributions to Order Statistics*. New York: Wiley.

Schwager, S.J. Mean slippage problems. In *Encyclopedia of Statistical Sciences*, Vol. 5, edited by S. Kotz and N.L. Johnson. New York: Wiley, 1985. pp. 372–375.

Somerville, P.N. (1954). Some problems of optimum sampling. *Biometrika*, **41**, 420–429.

Somerville, P.N. (1970). Optimum sample size for a problem in choosing the population with the largest mean. *Journal of the American Statistical Association*, **65**, 763–775.

Srivastava, M.S. (1973). The performance of a sequential procedure for a slippage problem. *Journal of the Royal Statistical Society, Series B*, **35**, 97–103.

Stein, C. (1981). The variance of the number of square-free numbers in a random interval. Technical Report, Department of Statistics, Stanford University.

Thurstone, L.L. (1919). The learning curve equation. *Psychological Monographs*, **26**, No. 3.

Truax, D.R. (1953). An optimum slippage test for the variances of k normal distributions. *Annals of Mathematical Statistics*, **24**, 669–674.

Chapter 5

CONTRIBUTIONS TO METHODOLOGY AND APPLICATIONS

Stephen E. Fienberg

5.1 Introduction

An apocryphal story tells of the statistician who encountered a biologist at a cocktail party many years ago. After some polite chatter about the weather, the biologist asked: "Have you ever heard of the geneticist, Sir R.A. Fisher? A colleague once told me that he dabbled in statistics." Of course, when Fisher did his pioneering work, statistics as a field was in its infancy, and the biologist's confusion might have been understandable. Today, those who make extensive statistical contributions to other fields are rightly viewed as statisticians, and in this regard Fred Mosteller is, to many, *the statistician par excellence.*

Early in his career, Fred Mosteller was often viewed as a quantitative social scientist, not simply a statistician, and this perhaps says much about the impact his work had on that of psychologists, sociologists, and others. Yet, at the same time, Fred worked on various problems in mathematical statistics and statistical methodology, many of which grew out of applications. Fred has also made extensive contributions to the use of statistical methods in medicine and in public policy, as well as excursions into such topics as the law and weather modification. Thus, it is impossible to separate his methodological work from his work in applications.

No single chapter could possibly do justice to Fred Mosteller's contributions to statistical methodology and to the applications of those methodological contributions in such a variety of substantive fields. Fortunately, because so many of these contributions have taken the form of books or edited volumes that are reviewed later in this book, in Chapter 8, the present chapter need not attempt systematic coverage. Rather, we focus on selected contributions to illustrate several aspects of the ways in which Fred has approached statistical problems and their solution.

The chapter is organized roughly by problem area rather than chronologically. Those who know Fred's work will not be surprised to learn that he has rarely written only a single paper on a topic, and thus his papers often form convenient groupings. Such an approach has its failings, however, because some papers defy classification, and themes woven into the fabric of Fred's work can often best be seen across areas rather than within a single area of application or a statistical topic.

5.2 Public Opinion Polls and Other Survey Data

One of Fred's first major applications projects occurred during his graduate student days at Princeton, where he worked as a research associate in the Office of Public Opinion Research under Hadley Cantril. This work not only led to a series of chapters and appendices authored or co-authored by Fred in Cantril's book *Gauging Public Opinion* (B1), but it also began a career-long interest for Fred in sample survey methodology and its applications. The chapters report on the substantive applications, and the appendices provide technical details for methodology used.

In Chapter IV, Fred and Cantril focus on the role of batteries of questions to understand the breadth and intensity of opinion on issues that have no clear public focus. They compare four different methods for characterizing interventionist/noninterventionist opinion in the U. S. just prior to its entry into World War II, using a battery of questions on the issue, and contrast these with the results from a single question with a related intensity scale. One of these methods was developed in a separate paper in *Public Opinion Quarterly* (P1a) by Fred and Philip McCarthy. There they use responses to eight questions to estimate the proportions in the interventionist/isolationist/no-opinion categories via a three-stage procedure. First they divide the sample into two parts, and they use the first part to estimate $p_{ij|k}$, the probability that someone in category k answers question i in category j. To do so, they need to identify groups of respondents in category k, $k = 1, 2, 3$; here they finesse the apparent circularity by assuming that 6 or more interventionist responses determine the interventionist group, 6 or more isolationist responses determine the isolationist group, and no more than 3 responses in either category indicate no opinion. The results are $\{\hat{p}_{ij|k}\}$. Then they use the second part of the sample to estimate p_{ij}, the marginal probability of answering question i in category j. Finally they estimate the proportions of interest, p_1, p_2, and p_3, by solving:

$$
\begin{aligned}
p_1 \sum \hat{p}_{i1|1} + p_2 \sum \hat{p}_{i1|2} + p_3 \sum \hat{p}_{i1|3} &= \sum \hat{p}_{i1}, \\
p_1 \sum \hat{p}_{i2|1} + p_2 \sum \hat{p}_{i2|2} + p_3 \sum \hat{p}_{i2|3} &= \sum \hat{p}_{i2}, \qquad (5.1) \\
p_1 \sum \hat{p}_{i3|1} + p_2 \sum \hat{p}_{i3|2} + p_3 \sum \hat{p}_{i3|3} &= \sum \hat{p}_{i3}.
\end{aligned}
$$

Although this approach appears somewhat ad hoc, Fred and McCarthy explore its robustness.

In Chapter VII, Fred investigates the reliability of basic background data gathered from respondents using repeat interviews by the same and by different interviewers. In Chapter XIV, Fred and Frederick Williams explore education and economic status as determinants of opinion on various topics. They use a specially constructed index, $W_{\text{ed}}/W_{\text{econ}}$, to determine the relative importance of the two background variables, where W_{ed} is a sum-of-squares-like quantity that compares counts of positive responses with those that would be "expected" if attitude were independent of education given economic status, and similarly for W_{econ}:

$$W_{\text{ed}} = \sum_{ij}(x_{ij1} - x_{i+1}x_{ij+}/x_{i++})^2 \tag{5.2}$$

where x_{ijk} is the number of respondents with economic status i and educational status j who respond k to the attitude question ($k = 1$ corresponds to a positive response).

Finally, in Appendix II, Fred describes a way to set limits on the errors due to interviewer bias under various assumptions, and Chapter VIII uses these in an actual study of interviewer bias. This was Fred's first involvement with nonsampling errors in survey work. The other appendices include material on stratification and pooling of data and on the use of charts to determine confidence limits for individual binomial proportions and differences in proportions.

Fred returned to sampling problems again in his evaluation of the pre-election polls for the 1948 presidential election (B3) and in his evaluation of the Kinsey Report (Kinsey, Pomeroy, and Martin, 1948) with William Cochran and John Tukey (B4). The main text of the Kinsey Report evaluation and a key appendix on the principles of sampling appeared as papers in *JASA* in 1953 and 1954 (P33, P34). The Kinsey Report evaluation is a fascinating document that is often cited as damning by Kinsey's critics and favorable by his supporters. In their introduction, Cochran, Mosteller, and Tukey (CMT) anticipated these extreme interpretations and noted that "We have not written either of these extreme reports." The CMT evaluation as a whole and the introduction in particular bear the imprint of Fred's approach to critiquing the work of others. Fred has often cautioned colleagues, roughly as follows: "Begin your criticism with a compliment. Remember this fellow has a lot of time invested in this enterprise, and if you are going to get him to change what he is doing, you need to convince him that you're on his side." The evaluation consistently tries to put the Kinsey work in the best possible light and not to hold it to unreasonable standards.

The CMT Kinsey evaluation addresses a number of crucial statistical issues: the selection of the sample and the inference to the target population, the accuracy of the interview data and possible methodological checks, techniques used in the analysis of the data, and the reporting of results. We focus here on the first two issues. On the matter of sampling, CMT note that Kinsey used "no detectable semblance of probability sampling ideas," and thus it is difficult to specify the nature of the sampled population as well as the gap between the sampled population and the target population. They then go on to describe how a small probability sample might be used "(1) to act as a check on the large [non-probability] sample, and (2) possibly serve as a basis for adjusting the results of the large sample."

Thirty-five years after the CMT evaluation, the U.S. government is considering the funding of a large-scale survey of sexual behavior, at least in part in response to the AIDS epidemic (see the related discussion in Turner et al. 1989). (Critics have objected to this survey on moral grounds, as well as because of its use of potentially inaccurate reports of behavior.) CMT caution us on the limitations of such surveys:

> In our opinion, no sex study of a broad human population can expect to present incidence data for reported behavior that are *known* to be correct to within a few percentage points. (p. 675)

but they also suggest that this still may be the best form of data available:

> Until new methods are found, we believe that no sex study of incidence or frequency in large human populations can hope to measure anything but reported behavior. It may be possible to obtain observed or recorded behavior for certain groups, but no suggestions have been made by KPM [Kinsey, Pomeroy, and Martin], the critics, or this committee which would make it feasible to study observed or recorded behavior for a large human population. (pp. 675–676)

They also suggest some methodological checks on the reported behavior that might strengthen one's belief in the accuracy of the data. These include (i) reinterviews with the same respondent, (ii) comparison of spouses, (iii) comparison of interviewers on the same population segment, and (iv) duplicate interviews by the same interviewer at various times.

On the nature of the interview, CMT foreshadowed a relatively recent statistical-cognitive movement to reexamine the form and structure of the survey interview (Jabine et al. 1984) by not accepting the primacy of the standardized survey questionnaire:

> The committee members do not profess authoritative knowledge of interviewing techniques. Nevertheless, the method by which the data were obtained cannot be regarded as outside the scope of the statistical aspects of the research.
>
> For what our opinion is worth, we agree with KPM that a written questionnaire could not have replaced the interview for the broad population contemplated in this study. The questionnaire

> would not allow flexibility which seems to us necessary in the use of language, in varying the order of questions, in assisting the respondent, in following up particular topics and in dealing with persons of varying degrees of literacy. This is not to imply that the anonymous questionnaire is inherently less accurate than the interview, or that it could not be used fruitfully with certain groups of respondents and certain topics. So far as we are aware, not enough information is available to reach a verdict on these points. (p. 693)

The separate CMT paper on the "Principles of Sampling" (P34) is a discursive review of some basic ideas in sampling with few formulas and much wisdom. The following is a non-random sample of comments and advice extracted from the paper:

> There are many ways to draw samples such that each individual or sampling unit in the population has an equal chance of appearing in the sample.
>
> Another principle which ought not to need recalling is this: By sampling we can learn only about the collective properties of populations, not about properties of individuals.
>
> Representation is not, and should not be, in any particular sample. It is, and should be, in the sampling *plan*.
>
> The ability to assess stability fairly is as important as the ability to represent the population fairly. ... One of the simplest ways [to do so] is to build up the sample from a number of independent subsamples ....
>
> The third great virtue of probability sampling is the relative definiteness of the sampled population.

This paper would still be useful required reading for all students of sampling before they begin to delve into the technical details of the subject.

Fifteen years later, Fred prepared a review of nonsampling errors for the *International Encyclopedia of the Social Sciences* (P74), and in it he picked up on many of the themes in "Principles of Sampling" and the Kinsey report evaluation. The review focused on nonsampling error in sample surveys and on techniques for reducing the bias associated with such errors. He begins with a general discussion of the conceptual error resulting from the differences between the sampled population and target population and then goes on to consider such matters as incompatibility of meaning, pilot studies, Hawthorne effects, and treatment integrity. He then turns to an overview of nonresponse and response errors in surveys, especially those involving personal interviews. On the matter of internal assessments of variability, Fred recalls the method of interpenetrating samples and goes on to describe the jackknife procedure and its properties of bias reduction and variance assessment.

A little over a decade later, Fred returned to a particular topic in sampling, the meaning of representativeness, in a series of four papers (P118 to P121) with William Kruskal. The first deals with uses of the term *representative sample* in nonscientific literature, the second with uses in the nonstatistical scientific literature, and the third with uses in the statistical literature. The fourth paper deals with historical developments. Kruskal and Mosteller catalog nine distinct usages of the term "representative sample," only a few of which can be found readily in the statistical literature. It is interesting to contrast these varied usages with the notion expressed in "Principles of Sampling" that representativeness is not a property of the specific sample but of the sampling plan.

5.3 Quality Control

In contrast with the continuity of Fred's work in the area of sample surveys, his work in quality control is concentrated in four clusters of papers that were written in the 1940s. The first is a cluster of size 1, dealing with runs in quality control charts, and was Fred's first paper (P1). In it Fred applies ideas from the recently-developed distribution theory of runs of k kinds of elements to quality control charts. He derives the probability of observing at least one run of length s, either above or below the median, and uses this probability to calculate the smallest values of s corresponding to $p = .05$ and $p = .01$ for the null hypothesis that a sample of size $2n$ is drawn independently from a continuous distribution function $f(x)$. The paper contains the first of many numerical tables he prepared.

The second cluster consists of material on sampling inspection plans in the volume (B2) that Fred helped to assemble (with H.A. Freeman, Milton Friedman, and W. Allen Wallis), growing out of the efforts of the Statistical Research Group at Columbia during World War II, led by W. Allen Wallis (see Wallis's review of this book in Chapter 8). The volume describes single-, double-, and sequential-sampling plans for acceptance sampling and discusses their properties, principally in terms of their operating characteristic curves (relating what we now refer to as type I and type II errors). The authors focus on the actual implementation of these plans, and they give considerable attention to details of random sampling in actual manufacturing settings. One of the two chapters on which Fred was the principal coauthor (Chapter 6) is devoted to the relationship between the inspection sampling schemes and the more traditional control chart methodology.

As a result of these wartime activities, Fred became a member of the editorial board of *Industrial Quality Control*, the bimonthly journal of the American Society for Quality Control. In 1949–1950 he wrote a four-part series of review articles (P10 to P13) with John Tukey based on material they presented at the Third Annual ASQC Convention at Boston in May 1949. The methods described in these papers range from "quick and dirty" methods for estimating means and standard deviations to methods for counted data, including transformations, to topics in the analysis of variance. The guiding philosophy throughout the review is that methods from various areas of application are relevant to problems in quality control. As Mosteller and Tukey note in the preface to the first paper, "Quality control people have been quick to apply the latest methods which have been brought to their attention, and they frequently ask for and need new and different tools to help solve their more specialized problems." The final paper in the *IQC* series returns to results more closely linked to the quality domain, including sampling inspection plans for continuous process control and advances in the design of experiments.

The final quality control cluster is also of size 1—a 1946 paper with M.A. Girshick and Jimmie Savage (P3), which presents a unified methodology for deriving unbiased estimates for Bernoulli sampling schemes and applies it to some standard QC sampling plans (e.g., single, curtailed single, curtailed double, and sequential sampling). The basic idea is simple yet ingenious. They consider a grid of integer-valued points in the xy-plane and let the point (x, y) correspond to y successes and x failures. The probability associated with any single path from (0,0) to (x, y) is $p^y q^x$, where $q = 1 - p$ and p is the probability of success. The probability of a boundary point for a sampling plan or an accessible point α being included in a path from the origin is

$$P(\alpha) = k(\alpha) p^y q^x,$$

where $k(\alpha)$ is the number of paths from the origin to the point α.

To get unbiased estimates of p, the authors consider the number of paths from (0,1) to the point α and denote it by $k^*(\alpha)$. Then they show that

$$\hat{p}(\alpha) = \frac{k^*(\alpha)}{k(\alpha)}$$

is an unbiased estimate of p. They note in passing that the same method yields unbiased estimates of functions of the form $p^t q^u$ for an accessible point (u, t) if (u, t) plays the role of (0,1). This approach yields unbiased estimates of variances and related quantities.

The remainder of the Girshick, Mosteller, and Savage paper deals with necessary and sufficient conditions for the uniqueness of unbiased estimates and then applies these results to the various QC plans mentioned above. Toward the end of the paper a brief section discusses combining estimates across sampling experiments (e.g., various kinds of acceptance sampling plans) and the difficulty in choosing among competing estimates in such settings. This is the first discussion of meta-analysis ideas in Fred's writings, a topic to which we return later in this chapter.

5.4 On Pooling Data

In a lovely little 1948 paper in the *Journal of the American Statistical Association* (P7), Fred discusses estimating the mean of a random variable X_1 from independent samples of equal size, n, for the normally distributed variables X_1 and X_2, each with known variance σ^2, but with unknown means. That is, one has the sample means $\bar{x}_1$ and $\bar{x}_2$, respectively, and wishes to know whether to use $\bar{x}_1$ alone or to take some combination of $\bar{x}_1$ and $\bar{x}_2$. He begins by describing his approach as a variant on the ANOVA technique of pooling estimates of error following a preliminary test of significance. Thus one would use the estimator $\bar{x}_1$ if a test rejects the null hypothesis that the two means are equal, that is, if

$$\mid \bar{x}_1 - \bar{x}_2 \mid \sqrt{n/2} > k\sigma,$$

and use $(\bar{x}_1 + \bar{x}_2)/2$ otherwise. The resulting estimator can then be written as a weighted linear combination of the pooled and unpooled estimators in which the weights are indicator functions, and Fred examines the mean squared error of this weighted estimator for various values of k. Not surprisingly, the new estimator does well when the difference between the population means, $d = \mu_1 - \mu_2$, is relatively small compared with σ.

The particularly interesting part of the paper comes toward the end when Fred adopts a partially Bayesian approach by giving d a normal distribution with mean 0 and variance $a^2\sigma^2$. Then he determines the MLE of μ_1,

$$\hat{\mu}_1 = \frac{\bar{x}_1(1 + na^2) + \bar{x}_2}{2 + na^2} = w\bar{x}_1 + (1 - w)(\bar{x}_1 + \bar{x}_2)/2, \tag{5.3}$$

where

$$w = na^2/(2 + na^2). \tag{5.4}$$

The key result is that this estimator has mean squared error less than the unpooled estimator $\bar{x}_1$. An estimator that has the form of equation (5.3) is now described in the statistical literature as a shrinkage estimator (e.g., see Berger, 1988), and results of the form described in this paper are typically traced back to Stein (1956) or to Robbins (1951).

5.5 Thurstone-Mosteller Model for Paired Comparisons

In a trio of 1951 papers (P18, P19, P20) in *Psychometrika*, Fred developed methodology for the formal study of a paired-comparisons model given by Thurstone (1927) and described earlier by Fechner (1860). In his monograph on paired-comparison models, David (1988) refers to this as the Thurstone-Mosteller model.

The basic description of the model as given in P18 is:

1. There is a set of stimuli that can be located on a subjective continuum (a sensation scale, usually not having a measurable physical characteristic).

2. Each stimulus when presented to an individual gives rise to a sensation in the individual.

3. The distribution of sensations from a particular stimulus for a population of individuals is normal.

4. Stimuli are presented in pairs to an individual, thus giving rise to a sensation for each stimulus. The individual compares these sensations and reports which is greater.

5. It is possible for these paired sensations to be correlated.

6. Our task is to space the stimuli (the sensation means), except for a linear transformation.

If we let X_i and X_j be the sensations evoked in an individual by the ith and jth stimuli, respectively, Thurstone's case V (and the one considered by Mosteller) assumes that the X_i have equal variances and equal correlations; i.e.,

$$\begin{aligned} \mathrm{E}(X_i) &= S_i, \quad i = 1, 2, \ldots, k, \\ \mathrm{var}(X_i) &= \sigma^2, \quad i = 1, 2, \ldots, k, \\ \rho(X_i, X_j) = \rho_{ij} &= \rho, \quad i, j = 1, 2, \ldots, k. \end{aligned} \tag{5.5}$$

For $d_{ij} = X_i - X_j$, we have under model (5.5) that

$$\mathrm{var}(d_{ij}) = 2\sigma^2(1 - \rho). \tag{5.6}$$

Without loss of generality we can set $\mathrm{var}(d_{ij}) = 1$ because the S_i are determined only up to a linear transformation. Then

$$\begin{aligned} p_{ij} &= P(X_i > X_j) \\ &= \frac{1}{\sqrt{2\pi}} \int_0^\infty \exp\left\{-\frac{[u - (S_i - S_j)]^2}{2}\right\} du \\ &= \frac{1}{\sqrt{2\pi}} \int_{-(S_i - S_j)}^\infty \exp(-y^2/2) dy. \end{aligned} \tag{5.7}$$

Expression (5.7) is a special case of the paired-comparisons linear model

$$p_{ij} = H(S_i - S_j), \tag{5.8}$$

where $H(x)$ is a cumulative distribution function for a random variable symmetrically distributed about zero. Another well-known version of the linear model is

$$H(S_i - S_j) = \frac{1}{4}\int_{-(S_i - S_j)}^{\infty} \operatorname{sech}^2(y/2)dy, \tag{5.9}$$

which yields the Bradley-Terry model

$$p_{ij} = \pi_i/(\pi_i + \pi_j), \tag{5.10}$$

for $\pi_i > 0$, $i = 1, 2, \ldots, k$. David (1988) uses the term "linear" to describe the model in expression (5.8), because the objects can be represented by k points on a linear scale with arbitrary origin.

In P18 Mosteller assumes that estimates of the p_{ij} are available (e.g., from independent binomial samples of size n) and that we can use expression (5.7) to transform these estimates using the normal inverse c.d.f. into estimates for the differences of the S_i, i.e., the $D_{ij} = S_i - S_j$. (Under the Thurstone-Mosteller model $D_{ij} = E(d_{ij})$.) He lets p'_{ij} and D'_{ij} be these estimates. The problem is that the sampling variability in the p'_{ij} will produce D'_{ij} that are not necessarily consistent. Thus Fred proposes to choose estimates S'_i for the S_i to minimize

$$\sum_{ij}[D'_{ij} - (S'_i - S'_j)]^2. \tag{5.11}$$

These least-squares estimates turn out to be

$$S'_i = \sum_{j=1}^{k} D'_{j1}/k - \sum_{j=1}^{k} D'_{ji}/k, \tag{5.12}$$

for $i = 1, 2, \ldots, k$, where we have arbitrarily chosen $S'_1 = 0$. Expression (5.12) had appeared previously in the literature for the uncorrelated case (i.e., $\rho = 0$), but it was presented as an intuitive estimate.

Fred ends this first paper with a discussion of the problematic situation when p'_{ij} is close to 0 or 1, so that the transformed values are extremely large. He suggests the exclusion of comparisons (i, j) for which $D_{ij} > 2$. As an alternative he suggests replacing the minimization problem in (5.11) by one directly using the p_{ij} in expression (5.7).

In the second paper in the trio (P19), Fred addresses the problem of an aberrant variance; i.e., the constant variances in expression (5.5) are replaced by

$$\begin{aligned} \text{var}(X_i) &= \sigma^2, \quad i = 2, 3, \ldots, k, \\ \text{var}(X_1) &= \sigma_1^2, \quad \text{with } \sigma_1 \neq \sigma. \end{aligned} \tag{5.13}$$

He then shows that, for this revised setup, the weighted least squares estimates of the S_i can be represented after a linear transformation as

$$\begin{aligned} S_i^* &= S_i^{'}, \quad i = 2, 3, \ldots, k \\ S_1^* &= [k(1 - 1/\sigma_d)/(k - 1 + 1/\sigma_d)]\bar{S}^{'}, \end{aligned} \tag{5.14}$$

where the $S_i^{'}$ are the least squares values under the equal-variance model and $\sigma_d^2 = \sigma^2 + \sigma_1^2 - 2\rho\sigma\sigma_1$. Thus all of the S_i are properly spaced relative to one another except for S_1, even if we mistakenly use the equal-variance assumption. Moreover, if S_1 is centrally located relative to the other stimuli, the effect of an aberrant variance is small.

In the last paper of the trio (P20), Fred proposes the use of a chi-squared goodness-of-fit test,

$$X^2 = \sum_{i<j} \frac{(\theta_{ij}^{'} - \theta_{ij}^{''})^2}{821/n}, \tag{5.15}$$

where $\theta = \arcsin\sqrt{p}$ is an angle measured in degrees, the $p_{ij}^{'}$ are the observed proportions, and the $p_{ij}^{''}$ are the estimated proportions under the Thurstone-Mosteller model. (The denominator in expression (5.15), $821/n$, is the variance of the transform of a binomial proportion with sample size n.) He then proposes to compare X^2 with values from a chi-squared distribution with $(k-1)(k-2)/2$ degrees of freedom. Because the estimation procedure used is not fully efficient, the actual asymptotic distribution of X^2 may not be χ^2. David (1988, p. 72, pp. 79–82) discusses this and related problems.

Fred illustrates the proposed test using data on the results from baseball games played in the American League in 1948 (the first appearance in print of Fred's interest in sports statistics). The League had 8 teams then, and Fred gets $X^2 = 14.78$ with 21 d.f., a remarkably good fit. He goes on to suggest that the fit may be too good and that the apparent stability may be a result of the failure to separate games played at home from those played away. (Many years later Fienberg (1980) replicated these findings using 1975 American League results.) In his conclusions Fred notes that the basic approach and the use of expression (5.15) work for any paired-comparisons model.

Seven years later, in his presidential address to the Psychometric Society (P43), Fred returned to the paired-comparisons problem, and he explored the effect of choice of the funtion H in the linear model of expression

(5.8). Specifically he shows in the context of a numerical example that five different transformations yield scaled values that are very similar (once they are subjected to a linear transformation). On the other hand, there are systematic differences in the estimated proportions p''_{ij} that correspond to the different models or choices of H.

After Fred's original trio of papers appeared, the literature on paired comparisons expanded rapidly, and the popularity of the Bradley-Terry model of expression (5.10) ultimately surpassed that of the Thurstone-Mosteller model, in part because of the formal statistical properties of the former as a special case of a loglinear model (see Fienberg and Larntz, 1976). Yet David's (1988) survey of the paired-comparisons literature clearly shows the influence of Fred's work on the topic.

5.6 Measuring Utility

Stimulated by work of von Neumann and Morgenstern and a paper by Friedman and Savage (1948), Fred set out with Philip Nogee, then a graduate student in social psychology, to measure "the value to individuals of additional income" (P24). This was a typical Mosteller enterprise—while colleagues were debating whether utility could actually be measured, Fred went out and did the empirical work.

The Mosteller-Nogee experiment to measure utility involved two groups of subjects, Harvard undergraduates and National Guardsmen. Subjects were asked to accept or reject bets of the form

> If you beat the following poker-dice hand, you will win \$$A$; otherwise you will lose \$.05.

The idea was to find the value A such that the subject was indifferent between accepting and rejecting the bet. If we write $U(X)$ for the utility of X, then the assumption is that there exists a fixed probability p_0 for a given individual such that

$$p_0 U(A) + (1 - p_0) U(-.05) = U(0). \tag{5.16}$$

By arbitrarily assigning 0 utiles (the unit of utility) to rejection of the bet and -1 to the loss of \$.05, Mosteller and Nogee needed only to calculate p_0 in order to determine $U(A)$. In effect the experiments were designed to treat the utility curve as a dose-response curve and to identify "the abscissa value at the point where the curve crossed the 50 per cent participation line" as the subject's indifference offer.

The experiments themselves are complex, and the utility curves differed considerably among individuals. The two groups of subjects, however, differed in a clearly identifiable way. For the Harvard undergraduates, the utility curves rose less and less rapidly as the amount of money increased (classical diminishing returns), whereas for the National Guardsmen, the larger the amount of money, the steeper the utility curve.

Edwards (1978) describes the key criticism of the Mosteller-Nogee experiment as the role the probabilities play. The probability of beating the specified poker-dice hand is not easy to calculate; yet it is used in the expected utility equation of the form (5.16) to represent the decision processes carried out by naive subjects. Not surprisingly, Mosteller and Nogee had raised this problem themselves as an issue.

The Mosteller-Nogee paper is a classic in the utility measurement literature and represents the first careful attempt to do empirical measurement. It was followed by extensive empirical and theoretical work by a variety of others (see Edwards, 1978). Fred recently returned to the "intuitive" understanding of probability in his work on the quantification of verbal probability expressions (P156, P180, P182).

5.7 Measuring Pain

In 1947 Fred Mosteller began to work with Henry K. Beecher and his colleagues in the Anesthesia Laboratory at the Massachusetts General Hospital. This collaboration extended well into the 1960s and led to a series of papers on the measurement of pain and the evaluation of the effectiveness of analgesics and anesthetics. Fred brought to this collaboration a knowledge of recently developed methodology and a conviction regarding the importance of statistical experimentation. The papers exhibit statistical sophistication without letting methodology get in the way of substance.

The first paper in this series (P17) appeared in 1950, and it deals with the measurement of pathological pain as distinct from pain experimentally produced and measured. It reports on a study involving 38 patients during the first 30 hours following a major surgical procedure and suffering from severe pain. Each was given varying doses of the drug to be studied and morphine, alternately, as often as was necessary with a minimum interval between doses of one hour. The morphine was always administered in constant dose, 10 mg. per 150 lb. of body weight. The experimental control is within patient, and the study was double blind. The response variable was subjective and binary: "relief" or "no relief." The paper includes a variety of interesting statistical ideas, including approximating the dose-response curve with a straight line over the range of response, the treatment of missing data, and the importance of adequate control and appropriate randomization.

The next paper (P22) focuses on the long-term effects of a (hypnotic) barbiturate used to reduce pain. Again the paper illustrates a careful double-blind, before/after crossover design, complete with randomization and a placebo. The primary analysis involves the straightforward use of t-tests on four different psychologial measures, but then the authors also report on the use of Fisher's method for pooling significance levels, noting the possible lack of independence and its effects. This is yet another signal of Fred's early interest in meta-analysis issues.

In a brief and separately-authored article in *Biometrics* (P27), accompanying an article by Beecher, Fred discussed the analysis of two problems arising in the measurement of subjective response to drugs. His first problem involves the use of McNemar's test in 2×2 tables for measuring change; the second deals with placebo-reactors, who respond in large part to drug administration per se, and he illustrates it with an example drawn from his work with Beecher. Here he does two tests, one the McNemar test to determine whether, in two separate studies, aspirin is better than placebo and morphine-codeine is better than placebo. Then he reorganizes the data across the two studies and looks separately at those who reacted positively to the placebo and those who did not. A more detailed description of this second problem and the data set was given in a multiple-authored paper (P31), and this was followed by yet another multiple-authored paper (P35) focusing on "the placebo response."

Part of Fred's role in working with Beecher's group was to keep the anesthesiologists aware of statistical problems in their research and to suggest methods for solving them. As such he prepared an expository chapter for a 1957 monograph-like review by Beecher describing the statistical methods used by his Harvard collaborators (P40, P44).

Finally, almost a decade later in 1966, Fred coauthored with Beecher and three others another paper on pain—this time experimentally induced (P65). The paper reports on three different experiments with designs similar to the early pathological pain studies.

Fred's papers with Beecher were part of Beecher's pioneering efforts to introduce scientific methods into the study of pain and anesthesiology. This effort not only helped to set the standards for work in this area, but it also introduced Fred to a number of young anesthesiologists who later sought Fred's assistance and participation in the National Halothane Study (B24)—see the related discussion in Chapter 7.

5.8 The Analysis of Categorical Data

Fred's methodological work on categorical data problems includes much of his work in other areas already considered here—for example, quality control (P1, P3), sample surveys (P1a), paired comparisons (P18, P19, P20), and the measurement of pain (P27). But Fred also made important methodological contributions in a number of other papers on a variety of related themes in categorical data analysis.

In a 1949 paper with John Tukey (P8), Fred describes the uses of binomial probability paper, which they had designed and published with the Codex Book Company in 1946 (M8a). The graph paper was originally designed to facilitate the use of R.A. Fisher's arcsine transformation for proportions. According to Mosteller and Tukey, Fisher

> observed that the transformation
>
> $$\cos^2 \phi_i = \frac{n_i}{n}$$
>
> transformed the multinomial distribution with observed numbers $n_1, n_2, \cdots, n_k$ into direction angles $\phi_1, \phi_2, \cdots, \phi_k$ which were nearly normally distributed with individual variances nearly $1/4n$ (when the angles are measured in radians). Thus the point at a distance $\sqrt{n}$ from the origin and in the direction given by $\phi_1, \phi_2, \cdots, \phi_k$ is distributed on the $(k-1)$ dimensional sphere nearly normally, and with variance nearly 1/4 *independent of n and the true fractions $p_1, p_2, \cdots, p_k$ of the different classes in the population.* The rectangular coordinates of this point are $\sqrt{n_1}, \sqrt{n_2}, \cdots, \sqrt{n_k}$.

Mosteller and Tukey's binomial probability paper (BPP) is graduated with a square-root scale on both axes and grid lines at counts of 1(1)20(2)100(5)400(10)600 on the horizontal axis and 1(1)20(2)100(5)300 on the vertical axis. (The notation $A(a)B(b)C$ means: "From A to B by intervals of size a and thence from B to C by intervals of size b.")

The 1949 paper is divided into six parts. Part I describes how to use BPP for plotting binomial proportions and splits (i.e., the true relative probabilities), as well as the observed *paired counts*, $(n_1, n - n_1)$. A split is plotted as a straight line through the origin that passes through all paired counts whose coordinates correspond to it. Thus the 80–20 split line in Figure 1 passes through the paired counts (8,2), (20,5), (120,30), etc. We put "heads" or "successes" on the x-axis and "tails" or "failures" on the y-axis.

Other topics in Part I of the paper include using BPP for converting splits into percentages, and measuring variation (one standard deviation is almost exactly 5 mm). Part II puts these ideas to use for some specific statistical tests and confidence intervals for one observed quantity, and Part III explains how to use BPP to plan a binomial experiment. Part IV addresses the analysis of several paired counts, including the analysis of 2×2 tables and bioassay problems. Finally Parts V and VI contain mathematical supplements and reference tables. An updated version of material on BPP and its uses is included in Mosteller and Tukey's 1968 chapter on data analysis (P76) for the revised *Handbook of Social Psychology.*

In a brief 1950 paper, Freeman and Tukey introduced variants of the usual variance-stabilizing transformations for binomial and Poisson counts. In place of the arcsine transformation for the binomial, they proposed the use of

Figure 1

Showing Plotting (Paired Counts, Splits, 1/10 Scale), Measurement of Distances (Short and Long), Calculation of Percentages, and Crab Addition
(reproduced from Mosteller and Tukey (P8) with permission)

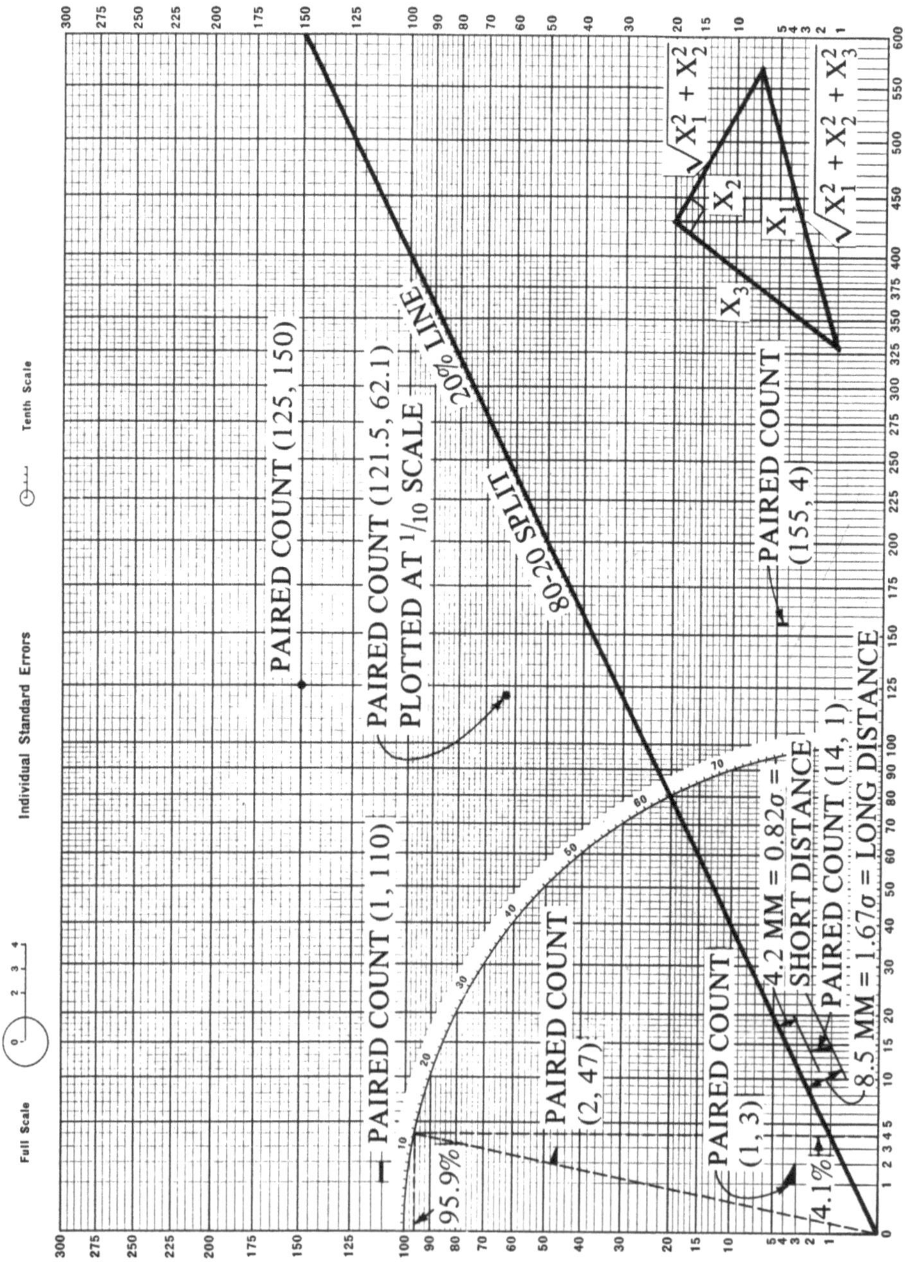

$$\theta = \frac{1}{2}\left\{\arcsin\sqrt{\frac{x}{n+1}} + \arcsin\sqrt{\frac{x+1}{n+1}}\right\}, \tag{5.17}$$

where x is the number of successes in n independent trials. In place of the usual square root transformation for the Poisson count x, they suggested the use of

$$g = \sqrt{x} + \sqrt{x+1}. \tag{5.18}$$

These variants were intended for small values of x or $n - x$. In 1961 Fred and Cleo Youtz (P50) published tables of θ (to two decimal places) for $n = 1(1)50$ and $x = 0(1)n$, and g ($x = 0(1)50$). In this paper they also explore, through the use of the tables, the dependence of

$$\sigma_\theta^2/\sigma_{\theta\infty}^2,$$

on p for various values of n, where $\sigma_{\theta\infty}^2$ is the "adjusted" asymptotic variance of θ, i.e., $821/(n+\frac{1}{2})$ when θ is measured in degrees. As n increases, the curve of variances tends to be quite flat for a long interval about $p = \frac{1}{2}$. By $n = 10$, we already can see the stability of the Freeman-Tukey variant compared with the original arcsine transformation.

In his 1967 presidential address to the American Statistical Association (P73), Fred took the occasion to report on a number of different categorical data problems that he was working on with various graduate students and colleagues. (See the discussion of the preparation of this paper in Chapters 6 and 7.) Most of the problems described in the published version of this address involve the use of cross-product ratios to measure association, and several were related to statistical work on the National Halothane Study.

In the Halothane Study Report (B24), Fred and Yvonne Bishop coauthored the key chapter on smoothed contingency table analysis. This chapter includes a discussion of loglinear models, and it applies them for each of five anesthetic groups for various strata of four population segments defined by risk and age. (An appendix by Bishop describes the use of the method of iterative proportional fitting to compute maximum likelihood estimates of expected cell counts under various loglinear models.)

Fred's interest in categorical data analysis continued, and he was the catalyst for the preparation for the comprehensive book *Discrete Multivariate Analysis: Theory and Practice* (B39) (see the discussion in Chapter 7).

5.9 Statistics and Sports

Fred Mosteller is a sports fan, and he has written three major papers on the statistical analysis of sports data—one about the baseball World Series (P28); a second about college football scores (P80), which led to two problem sets in *Statistics by Example* (B34, Set 7; B35, Set 11) and an abridged version of the paper (P108); and a third related paper on the use of resistant methods in the analysis of professional football scores (P113). In addition, he uses sports examples regularly in his textbooks and occasionally in papers (see the discussion of the baseball data in Section 5.5 on paired comparisons).

The World Series paper (P28), published in September 1952 just in time for that year's Series, is a vintage Mosteller paper. He begins with a practical problem—"Will the Series be very effective in identifying the better team?"—and formulates a simple probability model: independent binomial trials with a fixed probability of winning a single game, p, and truncated single sampling (i.e., the series continues until one team wins four games). Then he turns to available data and explores the superiority of the American League over the National League, using various estimators of p (including the one for truncated sampling from his paper with Girshick and Savage, P3). He does this for each series individually and then aggregates over time in different ways. The paper proceeds by examining various aspects of the original question and the binomial assumptions and the additional assumptions in the two different methods for deciding who is the better team coming into a given series. The paper ends with conclusions about the models used—they seem to be consistent with the data, and the two methods yield fairly close agreement. The abstract of the paper provides a summary of the findings, including an answer to the original question:

> The National League has been outclassed by the American League teams in a half-century of World Series competition. The American League has won about 58 percent of the games and 65 per cent of the Series. The probability that the better team wins the World Series is estimated as 0.80, and the American League is estimated to have the better team in about 75 per cent of the Series.

The discussion throughout the paper assumes a reasonable level of knowledge about baseball, how teams in the U.S. are organized, and even baseball lore (e.g., in the discussion of the possibility of serial correlation from game to game). These discussions and the paper as a whole represent one of the first major efforts to apply statistical modelling in a serious way to the area of sports.

In a 1970 paper (P80), Fred analyzes 1967 U.S. collegiate football scores in 1158 games, using informal, exploratory statistical methods to discover regularities. He begins by examining the joint frequency distribution of winning and losing scores; and, in contrast with the exposition in the World Series paper, he provides a careful discussion of scoring in college football. The "lumpiness" of the scoring system explains the high frequency of scores such as 14–7 and 7–0; multiples of 7 typically represent touchdowns with a one-point (kicking) conversion. Fred then turns to a series of issues: the chance that a given team score wins, predicting the winner's score from the loser's score and vice versa, the frequency distribution of winner's score minus loser's score. Interesting insights include the fact that ties are much less frequent than a well-fitting simple smooth curve suggests, whereas differences of 1 are much more frequent. (Fred offers two explanations, one based on an ad hoc probabilistic argument and the other on "the popular attitude of the players and the fans that a team should try to win rather than to tie.") Also, the sum of the scores, when plotted against the difference, is fitted well by the hyperbola

$$S = \sqrt{900 + d^2}. \tag{5.19}$$

The follow-up papers (B34, Set 7; B35, Set 11) explore these and other issues using both informal exploratory and more formal data analysis methods. It is interesting to note the absence of formal statistical models in this paper, especially in comparison with the World Series paper.

Fred returned to the analysis of football scores in a 1979 paper (P113), in which he uses resistant methods to analyze professional football scores. The key methodological tools in the paper are the stem-and-leaf display and the trimean (a weighted average of the first and third quartiles and the median). Separate analyses by team of "offensive" and "defensive" scoring make adjustments for the quality of the opposition. The concluding discussion is especially nice, and it returns to an early theme from the World Series paper: "What is the chance that the *best* of the eight teams ... entering the playoffs actually is the Super Bowl winner?"

Fred's interest in sports and statistics has spilled over into other writings. For example, the various general, elementary textbooks he has coauthored make extensive use of sports examples and problems. The indices in three such books contain the following entries:

> *Probability with Statistical Applications* (B11): baseball, handedness in baseball, World Series exercises.
>
> *Sturdy Statistics* (B33a): bowling, home runs, men on base, World Cup soccer.
>
> *Beginning Statistics with Data Analysis* (B48): baseball scores, college football scores, handedness in baseball, Indianapolis 500 winning speeds, Olympic platform diving, winning times for Olympic metric mile.

5.10 The Jackknife

John Tukey proposed the jackknife as a general-purpose method to reduce the bias of an estimator while simultaneously obtaining an estimate of its variability, thereby creating relatively reliable confidence intervals. The method was announced in an abstract (Tukey, 1958), and its properties were described in a number of unpublished memoranda. Quenouille (1956) independently suggested the same approach. For the next decade, however, the literature gave little evidence that anyone else had picked up the method. In his 1968 encyclopedia article on nonsampling errors (P74), Fred helped to popularize the jackknife, linking it to the method of interpenetrating samples and other methods for direct assessment of variability; and in a chapter for the revised *Handbook of Social Psychology* with Tukey (P76), Fred extended the jackknife approach to incorporate the idea of cross-validation.

The basic idea of the jackknife is simple: Divide the data up into k nonoverlapping groups. Let $y_{(i)}$ be the result of computing an estimator on the portion of the sample that omits the ith group. Let y_{all} be the estimate using the entire sample, and define the pseudo-values

$$y_{*j} = ky_{\text{all}} - (k-1)y_{(j)}, \quad j = 1, 2, \ldots, k. \tag{5.20}$$

Then the jackknifed estimate is

$$y_{*} = [y_{*1} + y_{*2} + \ldots + y_{*k}]/k, \tag{5.21}$$

and an estimate of its variance is given by

$$s_*^2 = [\Sigma y_{*i}^2 - ky_*^2]/k(k-1). \tag{5.22}$$

If the usual estimate, y, for θ has a bias whose leading term is of order n^{-1}, then the jackknifed estimate effectively removes this first-order bias term.

The idea of leaving out one group at a time appears also in cross-validation, where we estimate the parameter(s) of interest on all the data omitting the ith group, and then apply that estimate to predict or estimate for the omitted group. In their 1968 *Handbook* chapter Mosteller and Tukey extend the jackknife idea to include the cross-validation approach. By omitting two groups at a time instead of one as in the usual jackknife approach, they develop a jackknife estimate and simultaneously apply it to predict the estimate for the extra omitted group. This can, of course, be done in a multiplicity of ways. The chapter includes a detailed application of this dual approach to data from the *Federalist* Papers that Fred had analyzed earlier with David Wallace (B18). This chapter had a remarkable impact, even though it appeared outside the usual statistical literature. It helped to stimulate theoretical work on the jackknife by several statisticians (e.g., see Efron, 1981; Hinkley, 1983), as well as extensions such as the bootstrap, and it was the starting point for a separate literature on cross-validatory assessment (e.g., Geisser, 1974; Stone, 1974; Bailey, Harding, and Smith, 1989).

As in other areas, Fred made sure to present these ideas to other audiences, and we see them reexposited in a 1971 paper (P87), as well as in the book with Tukey on *Data Analysis and Regression* (B42).

5.11 Meta-Analysis in the Social and Medical Sciences

Most of those who have worked with Fred Mosteller (especially his former students) know that he takes great pains not to be critical of others, especially in print. As is noted earlier in this chapter, later in Chapter 7, and in the book reviews of Chapter 8, Fred makes a point to find aspects of someone else's work to compliment. Even when he has negative remarks to make, they tend to read like helpful suggestions that will improve the author's work in the future. From this perspective it is interesting to examine Fred's work in meta-analysis—the combining of results from multiple sources.

We note two items before proceeding to examine the papers themselves. First, Fred has focused on the exposition of meta-analysis techniques (see Robert Rosenthal's discussion in the appendix to this chapter) and on actually carrying out meta-analyses, which are the focus of this section. Second, by comparison with the meta-analytic work of others, Fred's approach has tended to be qualitative, rather than quantitative.

In several different but related research enterprises, beginning in the late 1960s and early 1970s, Fred with various co-authors (e.g., John Gilbert and Richard Light) explored the importance of controlled field studies (a carefully crafted euphemism for randomized experiments) in the evaluation of social programs. In the area of education we point first to his 1972 paper with Gilbert in the Mosteller-Moynihan volume (B32, Chapter 8) and the chapter with Light and Winokur in the 1971 Report of the President's Commission on Federal Statistics (B26, Chapter 6). These papers can be viewed as precursors to Fred's work in assessing such evaluation efforts.

The first major meta-analysis Fred carried out was reported in a lengthy 1975 paper with Gilbert and Light on assessing the impact of social innovations (P99). In this meta-analysis the individual evaluations measured the impact of the innovation. The studies reviewed span large-scale social action programs and studies in medicine and mental illness. The paper begins with a clear and gentle, yet blunt message: "In our collection of well-evaluated innovations, we find few with marked positive effects. Even innovations that turned out to be especially valuable often had relatively small positive effects—gains of a few percent, for example, or larger gains for a small subgroup of the population treated." (pp. 39–40) But even here the authors point to a silver lining and offer advice for the future (p. 40):

> Because even small gains accumulated over time can sum to

> a considerable total, they may have valuable consequences for society. ... The empirical findings for the programs described here emphasize the frequent importance of detecting small effects and, since these are difficult to measure, the need for well-designed and well-executed evaluations. Many would agree that, where practical and feasible, randomized field trials are currently the technique of choice for evaluating the results of complex social innovations. Our data suggest that such careful evaluations may be needed much more often in the future if society is to reap the full benefits of its expenditures for new programs.

This statement lays down a key idea in the meta-analysis literature—restricting attention wherever feasible to randomized controlled field trials. Another restriction of equal import is the decision to consider only published studies, revealed in a following section. This restriction has a dual purpose; it ensures that the study is accessible to others (including those doing the meta-analysis) and has undergone some kind of assessment in the process of publication.

Gilbert, Light, and Mosteller (GLM) end up reporting on their evaluation of 29 evaluations: 9 social innovations, 8 socio-medical innovations, and 12 primarily medical ones. They use a five-point scale,

++ = very successful
\+ = somewhat positive effect (but perhaps not worth the price)
0 = not much of an effect
− = slightly harmful
−− = definitely harmful

and end up with the following summary of the 28 single-goal innovations:

	Rating					
Type of Innovation	−−	−	0	+	++	Total
Social			3	2	3	8
Socio-medical		1	4	2	1	8
Medical	1	1	6	2	2	12
	1	2	13	6	6	28

One needs to take care in reading this table and the authors' conclusions. For example, the original New Jersey Negative Income Tax Experiment showed that the effect of providing negative income tax income to the working poor is a relatively small reduction in work effort. Because many had feared that a negative income tax would create a large reduction, GLM rate the proposed innovation as a considerable success (++). It is also interesting to note that Harvard Project Physics received ratings for two goals: a (0) for cognitive gain and a (++) for palatability to the students. Here is Fred's hand at work. On the primary goal, cognitive gain, the program had little discernible effect; but the GLM assessment takes special note of another positive feature and uses it to make the point that we often evaluate programs with multiple goals.

One shortcoming of the GLM study was the way in which the studies were selected, and thus Fred set out with John Gilbert and Bucknam McPeek (P106; B40, Chapter 9) to look at such assessments in a more systematic fashion. Their study focuses on randomized controlled trials in surgery and anesthesia. By searching the MEDLARS data base of the National Library of Medicine they located papers (in English) published between 1964 and 1972 with ten or more patients in a group—36 randomized comparisons reported in 32 papers, 19 dealing with primary therapies (intended to cure disease) and 17 with secondary therapies (intended to prevent complications). They use a six-point evaluation scheme with the 0's broken into two groups: those innovations that were regarded as a success because they provided an alternative treatment for some patients (0+), and those that were more trouble or more expensive than the standard (0−). The results mirror in many ways those from the GLM paper:

	Rating						
Type of Innovation	−−	−	0−	0+	+	++	Total
Primary Treat.	1	3	7	2	5	1	19
Secondary Treat.	3	3	3	2	2	4	17
	4	6	10	4	7	5	36

The overall conclusion here is similar to that in GLM: "Innovations do not automatically turn out to be improvements—even in the hands of the most careful people." (P123) "... averaged across these studies, the patient in the control group does about as well as the patient in the experimental group. An overall prior chance of success of 50:50 for these therapies is compatible with the results obtained." (P146) Gilbert, McPeek, and Mosteller (GMM) also develop a more quantitative method for examining the performance of the innovation with the standard therapy, and they apply this approach to a subset of the studies.

In the same pair of papers, GMM study 11 nonrandomized trials and note that they produce nearly half ++'s, thus reaffirming an earlier report by Grace, Muench, and Chalmers (1966) that less well-controlled studies produce more enthusiasm for the innovation than do well-controlled ones.

In an earlier paper focusing on the same data base of studies in surgery and anesthesia, GMM (P102) examine if and how investigators measure the quality of life after the recovery phase of surgery. In particular, they report on studies of quality of life in bone, ulcer, cirrhosis, and cancer surgery. They found reporting on the quality of life to be sporadic and especially rare for long-term follow-up, but with substantial variation from area to area. GMM report no quantitative summaries in this paper.

Two key components of these meta-analytic studies are the focus on randomized experiments and the need for quality in the reporting of study results, both topics on which Fred subsequently expended much effort. The work with Gilbert, Light, and McPeek simply reinforced Fred's already strong preference for the randomized experiment as the comparative method of choice. Fred continued to write about randomized experiments in medicine for much of the next decade, how they should be conducted, special features and approaches worthy of note, and the reporting of experimental results (e.g., see P126, P127, P136, P146, P147, P150, P151, P166, P177, P178, B56).

Despite the qualitative focus of Fred's meta-analysis work in social science and medical areas, he maintained an interest in what Richard Light refers to as "summing up" areas of scientific literature, and he kept abreast of the more formal developments of a quantitative flavor referred to in the Rosenthal appendix to this chapter. Indeed, he contributed to it, in a somewhat peripheral way, with a pair of recent papers with John Tukey on combining results from different experiments of a replicated or parallel form (P140, P149). In these papers, they extend an approach, developed by Cochran (1954) and referred to as partial weighting, which begins with preliminary weights for the different estimates that are inversely proportional to the estimated variances, and then assigns the same weight to the one-half to two-thirds of the estimates with the smallest s_i^2 (using the mean of the preliminary weights). In the first paper, Mosteller and Tukey assume, as did Cochran, that a common quantity is to be estimated across the experiments. They take the sample variances, s_i^2, for the n estimators, compute $s_{(i)}^2 = s_i^2(\nu_i+1)$ where ν_i are the corresponding d.f., and construct the standardized ratios

$$s^2(i)/c_i, \tag{5.23}$$

where $c_i = c(i \mid n, \nu_i)$ is chosen to be approximately the median of the sampling distribution of the ith largest of n values sampled from $\chi^2_{\nu_i}/\nu_i$. The idea is that the $s^2(i)/c_i$ estimate a common σ^2 for those i to which σ^2 applies. Thus they group together, as a low group, all estimates with

$$s^2(i)/c_i < 2\min_j\{s^2(j)/c_j\}, \tag{5.24}$$

and then repeat the process. Next, they take the estimates not in the final low group and repeat the entire calculation. In the second paper they extend this approach to the case where the quantities being estimated from the different experiments differ. In the first paper, the new method is illustrated on data from experiments to measure the gravitational constant g, and in the second paper they reanalyze a series of agricultural experiments presented in a 1930s paper by Yates and Cochran.

One of the most recent papers listed in Fred's bibliography in Chapter 2 also deals with meta-analysis. In October 1986, Fred was one of the participants in a Committee on National Statistics workshop on "The Future of Meta-Analysis," and his formal summary remarks will appear in the volume produced from that effort (P179).

5.12 Summary

Fred Mosteller has worked and continues to work on an impressive variety of problems in the area of statistical methodology, and he has been involved in a host of efforts to apply innovative methodology to substantive problems in the physical, social, and biomedical sciences.

As we look back through this chapter and Fred's work described in it, we can see several themes that help to describe how Fred works on methodology and applications.

Those of us who have worked with Fred Mosteller may recognize the following guidelines:

> Don't try to solve the most general version of the problem until you understand it reasonably well. Begin with the simplest version or special case—a leading case.
>
> Quality statistical methodology gains strength from empirical exercises, simulations, and real examples.
>
> Good statistical insights can and should be reused. Ideas of value in one area of application are often of value in others as well.
>
> If you fail to solve a hard statistical problem, don't despair. You are likely to have different insight and greater statistical knowledge when you examine the problem again later from different perspectives.
>
> Principles of statistical experimentation and random sampling are extremely important for statistical practice, but one often has to settle for data collected in other ways. A good statistician needs to know the strengths and weaknesses of different approaches instead of being dogmatic about statistical purity.
>
> If measuring some concept or feature of a problem seems conceptually difficult, try something anyway. The actual process of measurement and data collection is almost always revealing.

These are guidelines that others would do well to follow.

Appendix: Frederick Mosteller, Social Science, and the Meta-Analytic Age

Robert Rosenthal

Formal meta-analytic statistical procedures emerged over half a century ago, and the biological sciences were the earliest beneficiaries. These data-summarizing procedures, however, were needed at least as much in the social sciences. When they finally did come to us, Fred Mosteller, more than anyone else, brought them.

When, in 1962, I joined the old Department of Social Relations at Harvard, Fred Mosteller was a member. He had already influenced my interest in meta-analysis, and I imposed upon him regularly for advice and consultation. With great patience and wisdom, he instructed me—a notably mathematics-free social scientist—in various aspects of statistics, including meta-analytic methods.

Poor Cumulation and the Pessimism of the Social Sciences

Many quarters of the social sciences harbor a chronic pessimistic feeling that, compared to the natural sciences, our progress has been painfully slow, if indeed any progress has occurred at all. One source of this pessimism is the problem of poor cumulation: the social sciences do not show the orderly progress and development evident in such older sciences as physics and chemistry. The newer work of the physical sciences is seen to build directly upon the older work of those sciences. The social sciences, on the other hand, seem nearly to be starting anew with each succeeding volume of our scientific journals.

Poor cumulation does not seem to be primarily due to lack of replication or to failure to recognize the need for replication. Indeed, clarion calls for further research frequently end our articles. It seems, rather, that we have been better at issuing such calls than at knowing what to do with the answers. In many areas of the social sciences we do have available the results of many studies that address essentially the same question. Our summaries of the results of these sets of studies, however, have not been nearly as informative as they might have been, either with respect to summarized significance levels or with respect to summarized effect sizes. Until recently, even the best reviews of research by the most sophisticated workers have rarely told us more about each study in a set of studies than the direction of the relationship between the variables investigated and whether a given p-value was attained.

Meta-Analysis: The Social Science Chapter

In 1954 Gardner Lindzey edited the famous *Handbook of Social Psychology*. Chapter 8 of that two-volume work was entitled "Selected Quantitative

Methods" and was written by Frederick Mosteller and Robert Bush (P36). Four pages of that chapter (pp. 328–331) may be seen as the first text on the topic of meta-analytic procedures in the social sciences. A quarter of a century later, a spate of longer texts on meta-analytic procedures cited their indebtedness to the seminal work of Mosteller and Bush (Cooper, 1984; Glass, McGaw, and Smith, 1981; Hedges and Olkin, 1985; Light and Pillemer, 1984; Rosenthal, 1980; 1984).

The Mosteller-Bush meta-analysis pages focus on combining tests of significance. They begin with a reminder of the importance of the independence of the results to be combined and illustrate the fallibility of intuitions suggesting that probability levels can be combined by simple multiplication. The heart of these few pages is the description of three methods of combining probability levels:

1. The Fisher method, which uses the fact that the obtained p can be transformed to $-2\log_e p$, which can be treated as a χ^2 with 2 df.

2. The Stouffer method, which requires our transforming obtained p's to standard normal deviates (z) and finding the combined z from $\sum z/\sqrt{k}$, where k is the number of independent z's obtained.

3. The t test of the mean z, which simply tests the difference between the mean z and zero using (a) the sampling variation of the obtained z's and (b) the number of z's (k) to compute the one-sample t.

The Fisher method had been known to some social scientists, but the other two methods were new to almost all of us.

Yet another important contribution of the Mosteller and Bush chapter was the discussion of the weighting of the results obtained in each study. Meta-analysts may want to weight each study by sample size, by its importance, by its experimental elegance, or by any other reasonable characteristic of the study. The Mosteller and Bush chapter made these weighting procedures generally available to social scientists. Some thirty years later, Fred Mosteller, this time with John Tukey, continued this work on the problems of weighting in the process of combining multiple results (P140, P149).

References

Bailey, R.A., Harding, S.A., and Smith, G.L. (1989). Cross-validation. In *Encyclopedia of Statistical Sciences*, Suppl. Vol., edited by S. Kotz and N.L. Johnson. New York: Wiley. pp. 39–44.

Berger, J.O. (1988). The Stein effect. In *Encyclopedia of Statistical Sciences*, Vol. 8, edited by S. Kotz and N.L. Johnson. New York: Wiley. pp. 757–761.

Cochran, W.G. (1954). The combination of estimates from different experiments. *Biometrics*, **10**, 101–129.

Cooper, H. (1984). *The Integrative Research Review: A Social Science Approach.* Beverly Hills, CA: Sage.

David, H.A. (1988). *The Method of Paired Comparisons.* 2nd Ed. London: Charles Griffin.

Edwards, W. (1978). Decision making: psychological aspects. In *International Encyclopedia of Statistics*, edited by W.H. Kruskal and J.M. Tanur. New York: Macmillan. pp. 107–116.

Efron, B. (1981). *The Jackknife, the Bootstrap, and Other Resampling Plans.* CBMS Monograph No. 38. Philadelphia: SIAM.

Fechner, G.T. (1860). *Elemente der Psychophysik.* Leipzig: Breitkopf und Härtel.

Fienberg, S.E. (1980). *Analysis of Cross-classified Categorical Data.* 2nd Ed. Cambridge, MA: MIT Press.

Fienberg, S.E. and Larntz, K. (1976). Log linear representation for paired and multiple comparisons models. *Biometrika*, **63**, 245–254.

Freeman, M.F. and Tukey, J.W. (1950). Transformations related to the angular and the square root. *Annals of Mathematical Statistics*, **21**, 607–611.

Friedman, M. and Savage, L.J. (1948). The utility analysis of choices involving risk. *Journal of Political Economy*, **56**, 279–304.

Geisser, S. (1974). A predictive approach to the random effect model. *Biometrika*, **61**, 101–107.

Glass, G.V., McGaw, B., and Smith, M.L. (1981). *Meta-analysis in Social Research.* Beverly Hills, CA: Sage.

Grace, N.D., Muench, H., and Chalmers, T.C. (1966). The present status of shunts for portal hypertension in cirrhosis. *Gastroenterology*, **50**, 684–691.

Hedges, L.V. and Olkin, I. (1985). *Statistical Methods for Meta-analysis.* New York: Academic Press.

Hinkley, D. (1983). Jackknife methods. In *Encyclopedia of Statistical Sciences*, Vol. 4, edited by S. Kotz and N.L. Johnson. New York: Wiley. pp. 280–287.

Jabine, T.B., Straf, M., Tanur, J.M., and Tourangeau, R., Eds. (1984). *Cognitive Aspects of Survey Methodology: Building a Bridge Between Disciplines.* Washington, DC: National Academy Press.

Kinsey, A.C., Pomeroy, W.B., and Martin, C.E. (1948). *Sexual Behavior in the Human Male.* Philadelphia: W.B. Saunders Co.

Light, R.J. and Pillemer, D.B. (1984). *Summing Up.* Cambridge, MA: Harvard University Press.

Quenouille, M.H. (1956). Notes on bias in estimation. *Biometrika*, **43**, 353–360.

Robbins, H. (1951). Asymptotically subminimax solutions of compound statistical decision problems. In *Proceedings of the Second Berkeley Symposium on Mathematical Statistics and Probability*, edited by J. Neyman. Berkeley: University of California Press. pp. 131–148.

Rosenthal, R. (Ed.). (1980). *New Directions for Methodology of Social and Behavioral Science: Quantitative Assessment of Research Domains* (No. 5). San Francisco: Jossey-Bass.

Rosenthal, R. (1984). *Meta-analytic Procedures for Social Research.* Beverly Hills, CA: Sage.

Stein, C. (1956). Inadmissibility of the usual estimator for the mean of a multivariate normal distribution. In *Proceedings of the Third Berkeley Symposium on Mathematical Statistics and Probability*, Vol. 1, edited by J. Neyman. Berkeley: University of California Press. pp. 197–206.

Stone, M. (1974). Cross-validatory choice and assessment of statistical predictions (with discussion). *Journal of the Royal Statistical Society, Series B*, **36**, 111–147.

Thurstone, L.L. (1927). Psychophysical analysis. *American Journal of Psychology*, **38**, 368–389.

Tukey, J.W. (1958). Bias and confidence in not-quite large samples [abstract]. *Annals of Mathematical Statistics*, **29**, 614.

Turner, C.F., Miller, H.G., and Moses, L.E., eds. (1989). *AIDS: Sexual Behavior and Intravenous Drug Use.* Washington, DC: National Academy Press.

Chapter 6

FRED AS EDUCATOR

Judith M. Tanur

In 1961 Frederick Mosteller appeared on NBC's *Continental Classroom* as its national television teacher of the course on probability and statistics. His perspective on this course appears in P54 and will be discussed below. But from my perspective the course was a Godsend. At that time I was a young mother of two living on a Marine Corps base (where my husband was stationed) in southern California. I had dropped out of graduate school (or stopped out, I hoped) three years earlier because I was pregnant with my first child. That is the way we did things in those days. By 1961 I had two small children, very little free time, a determination to study statistics, and *Continental Classroom*! One of the few times of the day that young children will leave their parents alone is before seven o'clock in the morning—hence every morning I did my calisthenics in front of the television set and learned statistics from Frederick Mosteller. Although not registered for credit for the course, I sent for the textbooks and faithfully did all the homework. I was in awe of Frederick Mosteller—as students often were of professors in those days—but in an exaggerated fashion because he was not only a professor, but a professor on television!

Imagine my delight and disorientation, then, four years later, when I found myself, as the staff editor for statistics on the *International Encyclopedia of the Social Sciences*, working one-on-one with "Fred" on his article "Nonsampling Errors," making suggestions, criticizing occasionally, and even calling him by his first name! That project led Fred to invite me to work with him and others editing *Statistics: A Guide to the Unknown*, and to work together on several other educational projects.

Although my own acquaintance with Fred began mostly in connection with matters educational, and has long continued with additional collaboration on educational projects, I know firsthand only a small fraction of Fred's activities as an educator in the broad sense of the word. Thus material for this chapter comes not only from my own experience but also from others' reminiscences and writings, as well as Fred's own comments in writing and in person.

Fred deals with precollege education—developing materials and advocating probability and statistics in the school curriculum; he is involved in face-to-face education—teaching courses at the university, working closely with generations of undergraduate and graduate students (and on *Continental*

Classroom he was face-to-face nationwide); he is interested in educating the general public as well as policymakers; and he teaches teachers, and he has had a very special teacher-student relationship with his children, Bill and Gale. He has participated in writing textbooks at all levels—for students in high school, college, and graduate school, for practitioners in various scientific disciplines, and for policymakers. Further, Fred is a statesman in the field of education. In several instances he has carried out an assessment of the needs in an area of education, and then put together a sensible project to fill those needs. He has worked on evaluating educational outcomes—and on evaluating education evaluation. And in an even broader sense Fred has studied the very processes of education and learning through his work on mathematical learning models.

In Fred's own mind these activities must be intertwined. Insights he gains in any single activity must carry over to others; questions that arise in the process of solving one problem must provide inspiration for research in another area; and—as I know from personal experience—people he comes to know working on one project are recruited into others. Nevertheless, to derive some understanding of Fred's wide-ranging interests and achievements as an educator, I shall try to look at activities separately. Although the order is surely arbitrary, the very apposition helps to demonstrate the common threads running through the varied activities.

6.1 Precollege Education—Advocacy and Action

In the 1950s Fred was a member of a Commission on Mathematics of the College Entrance Examination Board, a group charged to study and make recommendations for the improvement of the mathematics curriculum in American high schools. The group recommended that some basic statistical concepts—such as graphical methods and simple numerical summary statistics—should be part of mathematics in general education, even for students whose education would end with high school. For students planning to go to college, more thorough training in probability and statistics ought to be offered. Working with a group from the Commission, Fred was instrumental in producing a text that could be used for such high school courses, *Introductory Probability and Statistical Inference for Secondary Schools* (B6), which was published in 1957. A revised edition (B7) and a *Teachers' Notes and Answer Guide* (B8) followed in 1959.

Almost a decade later, Fred asked whether there had been progress in the introduction of material on statistics and probability in the high school curriculum. In what he describes as not a serious sampling study of the matter (P69) he wrote to the mathematics departments of a number of secondary schools in New Jersey. His report, presented at the 1967 annual meeting of the National Council of Teachers of Mathematics, found some progress but not as much as could be hoped. Many schools were teaching a one-semester course concentrating on probability with some including a little statistics. In typical Mosteller fashion, this talk includes useful instructional examples, along with an account of variability across schools in curriculum, preferences, and needs; and it makes recommendations for improvement of secondary school instruction in probability and statistics.

One of these recommendations called for "a sequence of articles that explain just what the value of probability and statistics is in the real world, outside the sphere of gambling and urn problems—and explain this in much more concrete terms than we have given in the past. Exactly how are these ideas used?" (p. 829). To accomplish this, he called for the formation of a committee to attack the problem of improving statistical education in the country's schools. Accordingly, in 1968 the Joint Committee on the Curriculum in Statistics and Probability of the American Statistical Association and the National Council of Teachers of Mathematics was formed with Fred as its chair, a position he held until 1972. The Committee set out with a two-pronged attack on the problem of introducing these materials into secondary schools. As Fred had advocated in his 1967 talk, the first prong was to prepare the way for acceptance and use of such materials by educating the parents of school children, school superintendents, principals, and board members, as well as teachers of mathematics and their supervisors, about how broadly statistical tools are applied, chronicling past accomplishments and current contributions that statistics makes to society.

The result of this attempt to prepare the ground for curriculum changes was the volume of essays, *Statistics: A Guide to the Unknown* (SAGTU), published in 1972 (B27), with a second edition in 1978, and a third edition in 1989. With success stories of statistics drawn from the biological, social, and physical sciences, from business, government, and industry, the volume found wide use as a supplementary text in college courses on statistics. The first two editions, though widely distributed, do not seem to have found their way into the hands of the general public for which they were originally intended. It is hard to know why—perhaps marketing approaches were wrong, perhaps a volume with "statistics" in its title is doomed to exile from the best-seller list. Nevertheless, perhaps a sort of "trickle down" effect will occur. As students who were exposed in college to the usefulness of statistics—rather than simply to a set of dry formulas—become more influential in their local communities and in educational circles, perhaps they will be receptive to the introduction of statistics and probability into the curriculum because they are aware of the contributions the disciplines make to society.

The second prong of the Joint Committee's work was to prepare textual materials useful for secondary school courses in statistics and probability. This effort took the form of a collection of examples of uses of statistics, roughly grouped by mathematical level, to be used to introduce statistical concepts into existing high school mathematics courses. Entitled *Statis-*

tics by Example, four paperback volumes (B34–B37) were published by Addison-Wesley in 1973. These were subtitled *Exploring Data, Weighing Chances, Detecting Patterns*, and *Finding Models*. The hope was that these materials would be freely used in classroom instruction and as the basis for other texts.

Fred has stepped down from the chairmanship of the Joint Committee, but he has continued his interest in its work. The Committee is currently working on a major project on quantitative literacy, not only producing materials for high school classes but also providing materials for teacher training. It has held several teacher-training workshops to train individuals to deliver in-service training to other teachers. Thus there is hope for a ripple effect in the production of secondary school teachers who are qualified (and feel qualified) to present materials on statistics and probability in their local classrooms. And Fred led the preparation of the third edition of *Statistics: A Guide to the Unknown*, which may, in this incarnation, accomplish somewhat more in preparing educational decision-makers to choose to include statistics and probability in secondary school curricula.

During the 1960s and early 1970s Fred proselytized for statistics and probability as part of precollege education before many audiences. As early as 1963, for example, he argued the case before a Conference on the Role of Applications in a Secondary School Mathematics Curriculum (P60), pointing out that the

> approach to probability through finite sample space gives the student challenging mathematics, a refreshing introduction to the wealth and fun of applied mathematics following arithmetic and geometry, and it capitalizes on algebraic tools, as well as improving the students' skill in manipulation. It adds a new area of mathematics to the students' kit, and our social scientists need variety of kinds of mathematics. Since the ideas of probability are different and new, students take a long time to become comfortable with them, so it is appropriate that they begin this process with the easier material of finite sample spaces before they must mix in the difficulties of calculus. Finally, the product has direct and immediate payoff in everyday applications in the social sciences: notions of sampling, randomization, expectations, variation, and their treatment and analysis are utterly routine and everyday in work in the social sciences—not esoteric or rare, and the students' secondary-school course can supply him with a set of usable tools that are not just valuable sometime or maybe, but now and surely.

Reports of the work of the ASA-NCTM Committee were given at conferences (e.g., P86), at local and national meetings of the National Council of Teachers of Mathematics and the Mathematical Association of America, and in professional journals (P78a, P78b, P89). Each address or article not only described the Committee's work and invited contributions to it, but included several real examples of material that were to appear in the Committee's publications (analysis of collegiate football scores, ratings of typewriters to illustrate a 2-way layout, the local frequency of prime numbers). Thus Fred enlivens the presentation, piques the reader's or hearer's interest, and puts ideas for useful instructional materials into circulation rapidly.

In his 1967 address recommending the formation of the ASA-NCTM Joint Committee (P69), Fred also recommended that NCTM adopt an idea originated by Lennart Råde for an international conference on the teaching of probability and statistics in the secondary schools. The purposes of such a conference, he suggested, would be (P69, p. 831)

1. to explain in detail why probability and statistics can and should be taught in the secondary schools,
2. to discuss what sorts of things need especially to be taught,
3. to discuss the place of probability and statistics in the larger framework of mathematics, including school, secondary school (including those not collegebound) and college mathematics so that the secondary teacher can see the larger picture,
4. to describe good ways to introduce important topics to the students,
5. to review steps needed in teacher training, and
6. to establish a basis for continuing international cooperation in this field.

Such a conference was held—indeed, several conferences took place. The first, organized by Råde, in 1969 in Carbondale, Illinois, produced a volume (Råde, 1970) that, as Fred points out (P93), contains many fine projects and programs useful to the teacher. Fred categorized these examples by type and by article in his review of the conference volume for the *Review of the ISI* (R21). ISI then held several roundtables on related subjects. One in 1970 in the Netherlands dealt primarily with new methods of teaching (Hunter, 1971); another in Los Angeles dealt with computing (ISI workshop, 1973); and a third, in Vienna, was chaired by Fred and dealt with teaching statistics on the secondary school level. In his introductory remarks to that conference (P93) Fred described the work that the Joint Committee had accomplished, and later in the conference William Kruskal (1975) described the Committee's future plans.

Fred's advocacy for precollege courses in statistics and probability continues. As recently as 1982 he accepted membership on a Commission on Precollege Education in Mathematics, Science and Technology for the National

Science Board. The report of the Commission, *Educating Americans for the 21st Century: A plan of action for improving mathematics, science and technology education for all American elementary and secondary students so that their achievement is the best in the world by 1995* (B45), starts out by saying, "The Nation that dramatically and boldly led the world into the age of technology is failing to provide its own children with the intellectual tools needed for the 21st century" (p. v). The report urges that "we must return to basics, but the 'basics' of the 2lst century are not only reading, writing and arithmetic. They include communication and higher problem-solving skills, and scientific and technological literacy—the *thinking* tools that allow us to understand the technological world around us." (p. v). In order to meet these goals the Commission recommended that instruction at the K-8 level be designed to achieve, among other goals, an understanding of elementary data analysis, elementary statistics, and probability (p. 94) and that at secondary schools, education should be designed to achieve outcomes that include "understanding of elementary statistics (data analysis, interpretation of tables, graphs, surveys, sampling)" (p. 96). To help ensure these outcomes, the Commission recommended that colleges and universities include entrance requirements of course work covering probability and statistics (p. xi).

6.2 Face-to-Face Teaching in Traditional and Nontraditional Settings

In the various university departments in which Fred has had connections, he has taught courses ranging from undergraduate introductory statistics through extremely specialized graduate and faculty seminars. He has been mentor to large numbers of students who have gone on to successful and influential careers in statistics and related disciplines, and he has taught face-to-face nationwide on *Continental Classroom.*

In a required course on statistics in the social sciences, at times a class containing over 200 people, Fred managed slowly, carefully, and sincerely to get across the fine points of statistics. Carefully choosing examples, talking casually, sometimes seated casually, he is remembered as a wonderful teacher.

Fred is famous for treating graduate students as junior colleagues early in their graduate careers. As we shall see below, they are taught to be teachers, but research is also pushed early—part of graduate education is having a paper written up for publication, or even better, actually published. Fred pushes students, gives them ideas, dumps data in their laps, and encourages them to go ahead and do something with those data. The end result is that, even though students are "too busy" to do such publication work, because they are expected to do it they do indeed get it done. Gudmund Iversen remembers data from some four million rolls of a die, sent to Fred by a member of the television audience for *Continental Classroom.* With Fred's urging, supplying references, and guidance, and with Iversen's persistence, this turned into a supervised but practically independent research experience and a very useful publication (P85).

Students working with Fred on various funded projects are expected to produce professional summaries of their work for annual reports; and indeed, when Fred was preparing his presidential address for the American Statistical Association, he called upon several of his students doing research on a series of interrelated topics in the analysis of categorical data to outline their research for incorporation in that address. He circulated the address itself to students, as well as to other colleagues, for comments before it was given—thus helping to socialize students into the collegial and collaborative mode of research that has been so fruitful for Fred, for his collaborators, and for the profession. The result was P73 on estimation and association in contingency tables, rich with examples and careful (but often informal) exposition, and one of the few ASA presidential addresses so rich in technical content. It was also the first citation in the statistical literature for most of the students involved.

In a memorial essay for Jimmie Savage (P130) Fred spoke of profiting from that association in learning the steps of research. Despite his disclaimer in the first paragraph, it seems to me that this is precisely what Fred teaches students to do, and so I quote this advice in full (p. 28):

> I have formulated into six rules some things I recognized about Jimmie's way of doing research in those early days. They may seem rather simple and transparent to you, but a little reflection will show that some of them are the opposite of what we teach students to do, and, indeed, the opposite of how much research is practiced.
>
> 1. As soon as a problem is stated, start right away to solve it; use simple examples.
> 2. Keep starting from first principles, explaining again and again just what it is you are trying to do.
> 3. Believe that this problem can be solved and that you will enjoy working it out.
> 4. Don't be too hampered by the original statement of the problem. Try other problems in its neighborhood; maybe there's a better problem than yours.
> 5. Work an hour or so on it frequently.
> 6. Talk about it; explain it to people.
>
> I don't think these rules will make anyone a great researcher, but they do get one started.

> I personally benefited enormously from my close association with Jimmie right at the beginning of my career. I learned how to get started and to get out of ruts. I benefited, too, from writing papers with him and from conversations and associations through the years. Since then many young men have similarly benefited from knowing Jimmie, and their training is part of his scientific legacy along with his writings. In our sorrow today, let us not overlook our good fortune in having so rich a bank of memories to share.

Even graduate classes are treated as collections of junior colleagues. Michael Schwartz remembers a course with Fred that used a text that was then in preparation and that included what were then non-standard topics, such as the jackknife and Bayesian inference (P76). Surely by then no amateur in the construction of educational materials, Fred was perhaps surprised by the puzzlement the class expressed over these and other matters. He was surely the consummate teacher when he explained that one of the most important parts of the class was to be criticism of approaches, examples, and explanations. When communication failed it was the exposition, rather than the students, that bore the heavy measure of blame. (In a parallel fashion, Schwartz remembers, if an example did not "work" in class to illustrate a concept, Fred always had another up his sleeve—and yet another if necessary. Such examples, even the ones that failed to communicate with all class members, are remembered as cogent and intrinsically interesting.) Schwartz remembers that the class seemed a collective endeavor—students' work was taken seriously as indicators of strengths and weaknesses of the pedagogy as well as of the students. If something went wrong, it was something for everyone to think about; if something went well, students and teacher shared the happiness.

Supervision of Teaching Fellows demonstrates another phase of Fred's commitment to teaching and to the teaching of teachers. When a teaching assistant is scheduled to give a lecture, besides the usual preparation and planning with Fred, that student can expect a phone call—often in the evening when Fred first gets home—inquiring "How did it go, how did it work out, what examples did you use?" and so forth. Fred also models techniques for exam preparation and grading. Teaching Fellows submit trial questions, which are then worked over by Fred to make them more useful, and then—in a large class—all the Teaching Fellows meet with Fred to agree on the content of the examination. Similar cooperative work takes place in grading examinations. Fred considers each problem, talks about all possible errors and the misunderstandings they must reflect, and how much such misunderstandings should count against the student. Essentially, he creates an entire code book for grading before an exam is given. Surely it is easier simply to make up an exam oneself (or ask a Teaching Fellow to do so without much help) and then dump the exam in the Teaching

Fellow's lap for grading. But surely Fred's method makes the student work harder, helps the student learn more, and ends up creating committed professionals, encouraged to model themselves on Fred himself. (At least one former student reports using these same techniques with his own large introductory class and Teaching Fellows.)

Doing all this takes enormous amounts of time. Former students remark on Fred's apparently miraculous ability to stretch the day to something more than 24 hours. Not only does he get everything done, but he seems to do so effortlessly. When he works with a student, it is with regularly scheduled appointments, often held at Fred's home, in which Fred seems to act as if he had all the time in the world, almost never giving the impression of being rushed.

In the Statistics Department at Harvard, Fred felt that students should be involved in academic life in an even broader sense. Students were included in invitations to faculty cocktail parties (or, as one student described it, to parties with "adult people"), but they were invited one at a time. These were not mass "graduate student" parties, but opportunities for students to meet senior professors socially and to be introduced to the informal side of academic life. The more formal side was also part of the students' responsibility—when colloquia brought colleagues from other institutions to visit Harvard, Fred felt that students should be closely involved. Indeed a committee of students helped run the colloquium series.

According to several of his former graduate students, writing a dissertation under Fred is a learning experience not only about statistics but about how to write. Fred has a concept of a "zeroth draft" and comments extensively on such early drafts. Perhaps this way of working over text stems from an early experience that Fred describes in his tribute to Jimmie Savage (P130, pp. 25–26):

> Jimmie and I wrote a long paper together for the Statistical Research Group of Columbia, and had a lot of fun doing it. We were also rather pleased with ourselves for the way we had written it. Allen Wallis gave our manuscript to the executive director of the project, Milton Friedman, then a statistician, but best known today as an economist. Milton took the paper home over the weekend and sent it down to us on Monday. We could scarcely believe what he had done to the manuscript. Hardly a line was unchanged. The pages were black with corrections, and the backs full of extended notes and rewriting.
>
> Jimmie and I held an indignation meeting of the "He can't do that to us" variety, and we studied the literally hundreds of suggestions and changes that had been made. The more we worked, the more we found that Friedman was occasionally, well often, let's face it, usually right. We finally made a list of 100 or so objections to Milton's changes and went up seven flights

> to do battle with him. Milton was just delighted to see us. In no time he showed us that he was right in about 85 percent of the objections. He gave us credit for one or two complaints, admitted that one or two items were matters of taste, and the rest we agreed to disagree about. In the course of this discussion Milton kindly explained to us that we knew little about writing, that there were books from which we could and should learn, and he recommended several to us.

Fred's daughter, Gale, quotes this story and goes on to tell us how Fred profited from Friedman's advice

> by practicing his writing, exchanging editorial feedback with colleagues, and reading books about writing as well as books by good writers, my father developed his own rules of writing. He places passive verbs and the word "which" high on his list of writing scourges and relentlessly weeds them out.
>
> All through my school years, my father read through my drafted papers and reports and provided editorial feedback: page after page covered with markings, insertions, deletions, questions, clarifications, comments, and organization suggestions. For years, each page of my drafts had more of his writing than of mine because of all his editorial discussion. And I was not alone. He gave similarly detailed comments on papers written by his students and by his colleagues. Rather than merely pointing out that a sentence had poor wording or structure, he would generally make alterations on the sentence to show one way of improving the sentence. This approach helped the writer to become aware of the nature of the reader's confusion and to consider alternative wordings, even if the writer didn't use the modified sentence in the next draft.
>
> The next draft? A strong believer in writing as an iterative process, he repeatedly told me (and his students) to write on every other line of the page, leaving room for changes. He likes to call the first stab at a paper "the zero draft" to emphasize its unpolished nature.

Fred's son, Bill, echoes his sister's description of their father's insistence on re-writing, and extends that lesson to his field of systems programming.

> Growing up, I did not initially realize how much my father knows about writing. However, more through patient repetition on his part than receptive listening on mine, I have learned an important concept about writing. 'The first draft is a major milestone, but there is a long way yet to go.'

> The times I have produced what I consider good writing have always been associated with many reviews and major rewrites. I consider myself lucky that I understand that those reviews and re-writes are intrinsic in good writing, absolutely necessary and utterly unavoidable. ...
>
> Although Dad took the Fortran class [with Bill], he never actually wrote any further Fortran programs that I know of. He did teach me a concept about computers which has stood me in good stead ever since. 'Distrust any result produced by a computer.' I learned to manually calculate results produced by my programs and compare them. I learned to print intermediate results in defense, to simplify determining where the calculation went off track. I learned not to announce conclusions from the first run that completes, and have never regretted that lesson.
>
> The testing approach I use today is based on that outlook. I generally assume that any program I write does not work, and test accordingly. I believe that my work is better for that attitude. Fortunately, his insistence on accuracy in computer programs was coupled with a willingness to invest the time necessary to make sure we both understood what was expected and what was being prepared, and this made him easy to work with.

Gale reminds us that writing and revising and rewriting can be difficult, and that Fred, aware of that difficulty, works with students to overcome it.

> Because the hard work of research and writing can sometimes discourage us all, my father encourages students and myself to begin critiques by listing the virtues of the paper under review and by giving the author credit for his accomplishments thus far. In addition, he makes a habit of punctuating each milestone in the long process of writing with good cheer to emphasize the progress made. 'Hooray for you—you've finished a section (or chapter or memo or draft)!' He even cheers himself on. His enthusiasm has helped to keep me going draft after draft.

It is not surprising, then, that students remember marginal notes that are often negative, and remember as well the joy of getting something right and having Fred write in the margin "atta boy." (I remember getting something right as we were preparing *Statistics: A Guide to the Unknown* and having Fred put a gold star on it for me—I once even achieved the status of "heroine of the empire.")

Perhaps the most ambitious face-to-face teaching experience of Fred's career so far was as television teacher of NBC's *Continental Classroom* course on probability and statistics in the series on Contemporary Mathematics, broadcast first in 1960–1961. Broadcast at 6:30 in the morning, the course was offered for credit by many colleges and universities. Supported by several corporations and the Ford Foundation, and coordinated by Learning Resources Institute, the program was a large cooperative affair—for example, the teachers were chosen with the help of an advisory committee appointed by the Conference Board of the Mathematical Sciences. According to Edward Stanley, then the director of the Office of Public Affairs at NBC, probability was not then much taught in colleges and universities, and thus part of the idea of putting such a course on the air was to encourage teachers to learn the material so that they could incorporate it into their curricula.

Using, in part, material adapted from a Learning Resources Institute release (1961), Fred himself described the audience for the 1961 program (P54):

> The lectures were directed primarily at students taking the course for credit in collegiate institutions, but high schools, local boards of education, industrial and vocational organizations, and the general public used and viewed the program. Among colleges and universities 325 participated with an enrollment of credit students believed to be over 5,000. Approximately 66 high schools or school districts had about 1,200 students participating. At least 9 city boards of education have offered in-service credit to their teachers for Contemporary Mathematics courses. Some industrial organizations set up formal classes in their plants, often with one of the employees acting as teacher. Others had informal groups or encouraged their employees to watch. Members of the American Society for Quality Control, American Orthopsychiatric Association, and American Statistical Association were informed through news articles in their journals and through mailings of the existence of the course.
>
> From letters and surveys there appears to be a large non-credit audience among the general public. It has been estimated that during a week about 1.2×10^6 different people viewed the lessons. One wonders how many owed their presence to the six o'clock feeding, or to dialing in a morning news program a bit early.

It would seem that there are several ways that young children influenced their parents to watch *Continental Classroom*! Edward Stanley points out, however, that the early hour of the broadcast was chosen especially to permit teachers to watch before leaving for school.

Thus Fred was teaching face-to-face not to a small seminar, not to a large class of 200 students, but to a group that, on any given day, probably numbered in the hundreds of thousands. As Stanley recalls, the prospect seemed somewhat daunting to Fred at first, but he quickly—though perhaps not explicitly—grasped that the technique for sounding natural when teaching on television is to act as if you were addressing a single person rather than lecturing to a group. (Stanley points out that this is in contrast to radio, where one must pretend to be addressing a large auditorium.) This grasp of technique, and the fact that Fred turned out to be photogenic, soon produced a comfortable atmosphere and a teacher who was back in his element. In a piece that has been reprinted several times (P54), Fred has described in detail the process of preparing and taping a broadcast. (This is a typical Mosteller sequence: do something—often something innovative, describe what you've done—including how it's done and what could have been done better, and in the process teach others to do it too, and perhaps draw some morals for the entire teaching enterprise.) The description here tells of developing a television lesson (a process taking perhaps 3 1/2 to 4 days), including "planning for small physical demonstrations, films, equipment, weeding out dud examples and finding better ones, and in rehearsals" (p. 23). The program was rich with visual demonstrations, and Fred was delighted that the television environment gave him more time to develop such demonstrations than he had ever had before—and he describes some of them, including the simulation of an epidemic by mixing and pouring colored ping pong balls. He was also delighted that IBM, RCA, and General Foods each prepared a very short (3 to 10 minute) film for the program, and he expressed the wish that more such very short films, each illustrating a single concept, were available. If they were, he believes, classroom teachers would use them frequently to enrich their classes without the reluctance usually shown to devoting large chunks of class time to longer films. He makes a plea for the production of more such films, perhaps foreshadowing his later interest in developing modular instructional materials that expressed itself in the production of *Statistics by Example* and his advisory role in the project for Undergraduate Mathematics and its Applications (UMAP).

Classroom teachers face problems of timing, both of filling up the class period and of avoiding running overtime. But television exacerbates these problems because the material for a full 50-minute lesson is ordinarily compressed to the 30-minute time slot and because one simply cannot keep students "after the bell rings." Fred tells of learning to pace himself, speaking and walking more slowly than usual and diverting his students with exhibits and demonstrations, so that the material would not flow too fast for comprehension. Rehearsals (three formal ones in addition to more informal rehearsals at home) helped to refine the timing, but at the actual taping session small examples often had to be dropped when the director signalled to speed up. And examples, on television as well as in his conventional classroom, are always an important part of Fred's teaching. Edward Stanley recalls many discussions in which Fred tried out ideas for possible examples on him and vice versa. Often, the examples were drawn from sports (e.g., What are the chances that a batter with two strikes on him will hit a home run?) or other areas with which many viewers would be familiar. The pedagogic principle is that students will be more interested in, and thus more easily able to follow, an example from a field with which they are already familar and which poses a question they may already have wondered about. A problem, of course, is that students coming from different cultures (e.g., foreign students in American universities or watching American television, minority students if the examples are drawn from majority culture, or until recently, females if the examples are drawn from sports) are often at a loss to understand an example that the instructor thinks is perfectly clear. And the solution, of course, is the one Fred routinely adopts—include example after example, drawn from many different fields, so that if a student does not follow one, he or she has a chance to grasp the next.

6.3 On Teaching Teachers

It has been falsely said that those who can, do; while those who can't, teach. Going a step further, surely those who *really* can't, teach teachers. If Fred's teaching had not already put the lie to that foolish aphorism, his contributions to the education of teachers surely would. He refuses to be pinned down to a "philosophy of education," but his respect for his students, his insistence on teaching at a level on which he can be understood, and his writings describing his teaching "tricks" for the use of others add up to a pragmatic approach to the process of education. Although we can easily read some of the works cited above as addressed to the techniques of teaching, perhaps Fred made his most explicit statement on these matters in a symposium at the American Statistical Association meeting in 1978 and then in *The American Statistician* (P122). The article discusses the five main components seen as appropriate for a lecture (though often only four can be encompassed in a single lecture) and gives a great deal of useful advice to the teacher of statistics—or of any other subject. The five components advocated are:

1. starting the class with a few words about a large-scale application (*Statistics: A Guide to the Unknown* gets a plug as a useful source for such applications).

2. gathering data in class for a physical demonstration and involving students in their analysis. Here we find some tips on how to make the demonstration work and what to do when it doesn't. "If you think of possible misunderstandings, do not rely on the fact that no one asks a question after you ask for questions. If no one asks, then say, 'All right, then I have a few,' and then go over the instructions by asking questions about them and calling on specific students to reply." And "The teacher who uses physical demonstrations must be prepared for them to foul up

and also to take such failures in good humor when they occur. Some students get fun out of a teacher's discomfiture, so be prepared both to be cheerful anyway and not to blame anyone, including yourself." (p. 12).

3. presenting a small illustration, a specific use of a technique, complete with numbers. ("Fresh handouts revive the student, if the number is not overwhelming.")

4. for presenting new material, using a technique, credited to Robert E.K. Rourke, of going from the particular to the general to the particular. The particular is the small illustration used in (3) to motivate the lesson and clarify the technique. This is followed by a general, perhaps abstract, treatment of the technique. And then the lesson is driven home by a new example, similar to the introductory one, but with a new fillip. Interestingly, in conversation with me Fred insists he is uncomfortable teaching a class (or even witnessing one taught) that uses the old formula "tell 'em what you're going to tell 'em; tell 'em; then tell 'em what you've told 'em." But this particular-general-particular formulation seems very similar—perhaps the difference is the students' involvement in the first "particular" and the additional information transmitted with the second "particular."

5. establishing "plausibility—after proving and applying a theorem spending some time seeing whether teacher and class can make the whole thing seem reasonable as well, often qualitatively."

Not mentioned in this article is another element of Fred's educational credo—"First you have to give them something before you start taking it away." In conversation he describes an instructor who introduces a new technique—say regression—to the class with many warnings about when it is not appropriate, what its shortcomings are, and how its results are to be carefully interpreted. Although all these cautions should indeed be expressed (perhaps a month later), at introduction to the technique the class knows nothing about its uses or virtues and can hardly be blamed for wondering "If it's all that rotten, why is he telling us about it?"

The remainder of the article gives "helpful hints" on lesson preparation and presentation. Rehearsal is crucially important—and not simply silent rehearsal that one can do in an armchair, reading one's notes. Fred advocates timed, out-loud, full "dress rehearsals" of demonstrations, of board uses, of calculations. And the rehearsal should produce action—if a demonstration takes a half-hour instead of the planned five minutes, it needs to be re-thought or abandoned; if a lecture runs long, it should not be shortened by talking faster, but the instructor needs to be prepared to drop chunks; if not enough material is available to fill the time, the instructor should plan to start right in and present what's most important with the thought that questions are likely to take up the spare minutes. Board writing is discussed—the advice is "write large"—as is the use of overhead projectors, with advice again to write large and use few lines to a page. Further "the speaker may not realize how much small movements are magnified by the projector. When pointing to something with a pencil, rest the tip on the plastic to stop the quivering, or lay it down. Don't point with fingers, and don't wave hands over the transparency. All of these sudden movements are hard on the viewers' eyes and make some viewers seasick" (p. 15).

Finally the article advises what to do when disaster strikes—at home or on the lecture circuit. Perhaps the advice can be summarized as "be prepared and take it as it comes with good humor." But the details are arresting, and the article repays repeated readings.

6.4 Evaluating Education and Evaluating Educational Evaluations

Prescribing what "ought to work" in education isn't enough—it is important to follow-up to see whether what is being taught is indeed being learned, and further to understand how definitive such evaluations actually are. For years Fred worked on the National Assessment of Educational Progress, a project of the Education Commission of the States that administered standardized tests to students across the country and presented the results in a large number of reports.

Questions about how well schools and their staffs perform (as well as how students perform) have also preoccupied Fred. For a classic study of the organizational effects and determinants of variation in the performance of administrators in schools (Gross and Herriott, 1965), Fred (together with Arthur Dempster) helped develop a scale to measure executive professional leadership and published an appendix to the study on the question of how to build a model to weight observers' ratings of an object or person (P62).

A similar psychometric scaling was used to suggest additional analyses of the effects of education in a review essay on *The Enduring Effects of Education* by Herbert Hyman, Charles Wright and John Reed (R22). Applauding the pathbreaking secondary analysis of some 55 surveys of the adult population, Fred and his collaborators make it clear that they have strong biases that encourage them to believe the work's conclusion that the effects of education are both broad and enduring. But they go much further. They explore the need for models of the education process to account for the effects of background variables, discuss problems of selection effects, and, after pointing out some possible shortcomings in the methods the authors use to summarize and combine the results of the many surveys, propose a psychometric scaling model and demonstrate the virtues of an analysis using such a model. A good deal more than a book review is produced.

Perhaps Fred's most impressive contribution to evaluating educational evaluations is his co-editorship (with Daniel P. Moynihan) of *On Equality of Educational Opportunity* (B32), a set of papers arising from a Harvard Faculty Seminar in 1966–1967 which Moynihan (together with Thomas F. Pettigrew) co-chaired and in which Fred was clearly an enthusiastic participant. The overview chapter by the editors describes the findings of the "Coleman Report" (Coleman et al., 1966) and highlights the reanalyses in later chapters that re-examine some of the report's conclusions. This introductory chapter also sounds some of the themes so often found in writings that Fred participates in. We are reminded of the usefulness of even crude statistics in providing the feedback on public policy necessary for a nation to see whether it is making progress toward goals it had set, even if it has changed its mind somewhat about those goals. We are also reminded that findings that surprise us—in the sense of being contrary to our expectations—often invite disbelief and reanalysis, whereas findings that confirm our pre-conceived notions are rarely questioned. We are given brief explanations of the uses of the standard deviation, of the process of partitioning variances and of regression, and the dangers of statistical testing of post-hoc hypotheses—Mosteller and Moynihan are consummate teachers, on duty at all times. We are reminded of the importance of mathematics education for occupational choice and the ability to communicate. And we are given a careful policy analysis of measures needed in any attempt to close the educational gap between minority and majority groups, with stress not only on innovation but also on research and assessment and on an attitude among the public that expects small gains and is pleased to find them.

6.5 The Learning Process—Research, Teaching, and Practice

In early work with Robert Bush within the reinforcement-learning tradition (e.g., B5) Fred contributed to an understanding of how mathematical models can depict the acquisition of knowledge. Later (P81) Fred used this material to illustrate the connections between the mathematical sciences and the social sciences. One key finding was that one rewarded trial produced about the same change in response probability as did three punished trials. Fred seems to practice what his research preaches—I now understand the gold stars and the "heroine of the empire" award.

But in a very profound sense, this blending of research and practice epitomizes Fred's style as a teacher and as a scientist. The body of this chapter illustrates repeatedly how he draws research examples into his teaching;

how he trains students simultaneously in teaching and research; how his advocacy for better statistics education extends to those in practice (whether as teachers, as researchers, or as policymakers—e.g., B44). The threads are impossible to disentangle, for they are woven together into the very fabric of what makes Fred an outstanding statistical educator—indeed, of what makes him a statistical model.

6.6 Continuing Education

When I visited Cambridge for the celebration in honor of Fred's 70th birthday, he was teaching Statistics 100, the introductory course for undergraduates, for the last time before his "official" retirement from teaching. One would think that such a time would be an occasion to "coast" and teach the course the way it had always been taught. But Fred told me about some ideas that he had garnered from a seminar he was attending of senior faculty and deans of colleges across New England, gathering monthly to talk about teaching techniques. It had been suggested that at the end of each class the instructor ask the students to write out what the key point of the lecture was and what the student wanted to know more about. Fred tried this out for several weeks and handed out a frequency distribution as feedback to the students, tabulating the results and correcting egregious errors in understanding the key points. He found that what the students wanted to know more about was essentially more about the same. Finding the technique surprisingly less useful than it had been advertised to be, Fred discussed it with the class. They said it gave them an opportunity to integrate the material they had learned. (Fred noted that, although he makes a practice of listing the key points of a lecture in the left-hand corner of the board, students were not consulting the board to make up their list of key points but leafing through their notes.) But a student pointed out that the reason this technique was not as useful as possible was because it was not really giving any feedback to Fred himself. It turned out that what was really missing was "negative" feedback—and Fred added to the list of things that students should report on: "What was the muddiest point in the lecture?" He found then that the students were asking to know more about points that would clear up the muddiness. Saying that it takes him approximately an hour to put together a handout answering the broad questions, and sometimes an extra handout carrying the discussion into more esoteric points for which some students have asked for elaboration, Fred expressed himself as delighted with the success of this teaching innovation—sufficiently delighted, I later learned, to write it up for publication (P173).

In another way Fred's educational activities continued strong into 1986. In that year he published (with John C. Bailar III) *Medical Uses of Statistics* (B56), which was once again the result of a group effort that explored the uses of statistics in the *New England Journal of Medicine* and then turned around and attempted to teach good use of statistics for biomedical researchers. The stress of the book is on the broad concepts of statistics rather that on cookbook uses. Fred's own article is entitled "Writing about Numbers," and he uses twenty pages to give good advice on how writers can present numbers so that they will be more understandable to readers. Here is Fred once again, learning, organizing teaching, teaching about writing, and editing a book to make it all available to a broad audience.

ACKNOWLEDGMENTS

Special thanks are due to many people who shared reminiscences about Fred with me in preparation for this chapter or who helped me to contact others who had direct knowledge and to those who read and commented upon early drafts of this chapter: Susan Ellenberg, Stephen Fienberg, David Hoaglin, Gudmund Iversen, Livii Joe, William Kruskal, Richard F. Link, S. James Press, Michael Schwartz, Albert Shulte, Edward Stanley, and Cleo Youtz.

References

Coleman, J.S., et al. (1966). *Equality of Educational Opportunity.* Washington, D.C.: U.S. Government Printing Office.

Gross, N. and Herriott, R.E., Eds. (1965). *Staff Leadership in Public Schools.* New York: John Wiley & Sons, Inc.

Hunter, J.S., Ed. (1971). New techniques of statistical teaching: Report of the International Statistical Institute Round Table, Oisterwijk, Netherlands, 9–12 September 1970. *Review of the International Statistical Institute*, **39**, 253–372.

ISI Workshop on Statistical Computation. (1973). *International Statistical Review*, **41**, 225–275.

Kruskal, W.H. (1975). Towards future activities of the JCCSP. In *Statistics at the School Level*, edited by Lennart Råde. Stockholm: Almqvist and Wiksell; New York: John Wiley & Sons, pp. 187–195. A report of the Proceedings of a conference on the teaching of statistics at the secondary school level, held at Vienna, Austria, August 30 – September 4, 1973.

Learning Resources Institute. (1961). Release appearing in *American Mathematical Monthly*, **68**, No. 7, 666–667.

Råde, L., Ed. (1970). *The Teaching of Probability and Statistics: Proceedings of the First CSMP International Conference.* Stockholm: Almqvist and Wiksell; New York: Wiley Interscience.

Chapter 7

FRED AT HARVARD

Edited by
Stephen E. Fienberg and **David C. Hoaglin**

with contributions and recollections from
John C. Bailar III, Donald M. Berwick, John P. Bunker,
Doris R. Entwisle, William B. Fairley, Harvey V. Fineberg,
Ray Hyman, Stephen W. Lagakos, Nan Laird,
Louis Lasagna, Richard J. Light, Thomas A. Louis,
Lincoln E. Moses, William Mosteller, T.E. Raghunathan,
Donald B. Rubin, Stephen Thomas, Leroy D. Vandam,
James H. Ware, Cleo Youtz, Marvin Zelen

7.1 Introduction

In his biography of Fred Mosteller (Chapter 1), John Tukey notes that, after receiving a Ph.D. from Princeton, Fred spent his entire professional career at Harvard University, and in the process chaired four different Harvard departments. Thus it is no surprise that many of Fred's research and educational projects are inextricably intertwined with people and activities that have had some connection with Harvard. As we reread the other chapters of this volume, we thought again of Fred's roles as an educator, organizer, and researcher at Harvard; and in the end we chose to stitch together a quilt-like story about these activities and their impact on other people.

This account is primarily chronological, although we have occasionally skipped around to present a coherent description of some sustained topics. Some topics overlap with material in other chapters, especially the book reviews since so many of Fred's activities have been organized around the preparation of special books and monographs.

7.2 In the Department of Social Relations

Fred Arrives at Harvard

On January 29, 1946 the Faculty of Arts and Sciences at Harvard voted to establish a Department of Social Relations. The impetus for forming the department and its research wing, the Laboratory of Social Relations, came from a distinguished group of faculty members in anthropology, psychology, and sociology. Samuel A. Stouffer became the director of the Laboratory and Talcott Parsons the Chairman of the Department. From the beginning, methodology was to play a key role. The Department brought Fred to Harvard from Princeton and appointed him as Lecturer in Social Relations and Research Associate in the Laboratory of Social Relations.

Fred recalls the series of contacts and projects during the Second World War that led to this appointment (P168, p. 138):

> [Sam Wilks] also got me a job working with Hadley Cantril. Hadley was a social psychologist who pioneered in using survey research for social science. And as a result of that effort, I ultimately met Sam Stouffer and worked as a sampling consultant to the War Department in Washington. Finally, that led to my coming to Harvard University in the Department of Social Relations after I completed my degree at Princeton.

In a report to the Faculty of Arts and Sciences on the occasion of the tenth anniversary of the founding of the Department, Talcott Parsons (1956) noted the importance of Fred's appointment to the newly formed Department of Social Relations and his subsequent contribution:

> That [appointment] of Dr. Mosteller was designed to bring to the Department from its very beginning the highest level of thinking and competence in this rapidly developing field of the methodology of science. (p. 17)
>
> As noted, Mosteller had been brought to Harvard with the opening of the Department, as a mathematically trained statistician who could make available to our staff and students the critically important new theory and techniques which had been developing in that field. This contribution proved so important that the Department took the unusual step of recommending that one of its first two rare and precious openings for permanent appointments should be devoted to this field, and unanimously recommended Mosteller for it. (p. 25)

Fred became Associate Professor of Mathematical Statistics in 1948, after just two years at Harvard (at that time Associate Professor was a tenured rank at Harvard). He was promoted to Professor of Mathematical Statistics in 1951, and was the acting chairman of the Department of Social Relations during 1953–1954.

Early Teaching in Social Relations and Mathematics

As a member of the Department of Social Relations, Fred taught a variety of courses, including "Quantitative Methods in Social Research" and "Mathematical Statistics" almost every spring term. He also collaborated with many social scientists in offering numerous seminar courses, with titles such as "Measurement in Social Psychology," "Measurement of Attitudes," and "Methods of Research in Social Relations," to name a few.

Although statistics as a field of study was not widely recognized at that time, Fred attracted a considerable number of students to his courses. For example, in the spring of 1951, 32 students enrolled for "Mathematical Statistics"; in the fall of 1952, 121 students enrolled for "Principles of Statistical Inference," offered by Mosteller, Phillip J. Rulon, and Guy H. Orcutt.

Ray Hyman recalls some of his experiences co-teaching with Fred.

> Fred and I shared the teaching of the statistics course for the graduate students in the Department of Social Relations. Fred attended all my lectures and I, of course, attended his. Fred was a superb teacher and took great pains to make sure that each student fully understood both his lecture material and his answers to their questions. The one thing that Fred could not tolerate was a student who interrupted the flow of his lecture by coming to class late. When a student did enter the classroom after he had begun his lecture, Fred would stop talking and stare at the student. When the student had finally settled into his or her seat, Fred would sarcastically ask if he could resume his lecture. I remember that Fred had a repertoire of other reactions to latecomers, but I cannot recall the specifics. At any rate, that student and all the others got the message. After one or two incidents, no one ever showed up late for a Mosteller lecture during the remainder of the course.
>
> Since I was a psychologist with very little formal training in mathematics and statistics, I am sure that in many of my lectures I made statements or answered questions in ways that were technically inaccurate. Fred must have winced at my gaffes; however, he was careful never to publicly question or criticize me.
>
> Instead, he would approach me after the lecture and casually comment, "You know, Ray, that was an interesting observation you made about partial regression coefficients [or whatever the specific topic had been]. Let's go up to my office and see if we can work out its implications." We would then closet ourselves

in Fred's office. Often he would produce a deck of cards or some dice, and we would work out a crude simulation of some of the consequences of the topic I had discussed in the lecture. Once we had some idea of how a distribution or a relation might look in a simplified empirical model, Fred would then have us attempt to work out the consequences in a more formal and mathematical way. As a result, I usually discovered that what I had said in my lecture was wrong, and I was able to provide a correction in my next lecture.

In all these instances, I am sure that Fred already knew the answer. Yet he always acted as if "we" had discovered the correct formulation together. I hope the students benefited from this pedagogical method that Fred used with me. I know I learned a great deal from it.

Fred seemed incapable of treating any question from a student as trivial or stupid. No matter what the student asked, Fred managed to find interesting implications. Often the teaching assistants and I would enter Fred's office to find him patiently explaining to a student the basics of such elementary ideas as the use of decimal points, the calculation of the mean, or even what the coordinates on a graph mean. Because the teaching assistants could easily handle such questions, we often suggested to Fred that he not waste his precious time with such matters. But he rejected our advice. If a student asked him a question, no matter what the topic, he felt it his duty to treat the question seriously and to provide the best possible answer.

Fred's willingness to find something worthwhile in every student's question or answer sometimes seemed to go to extremes. Fred, I, and some faculty members from Economics and Education team-taught a general course in statistics for graduate students. At the end of the semester we would gather in one place and grade the final examinations. After we had apparently finished grading all the examinations, Fred would insist that we carefully review each examination that had received a failing grade. Invariably, Fred would develop elaborate theoretical reasons why an answer that the rest of us had assigned a failing grade was a brilliant way of looking at the statistical problem. Because he had found a way to justify the seemingly inappropriate answer, Fred would argue for giving a higher grade to the paper.

Because I was the youngest member of the teaching team, I often knew the student on whose behalf Fred was making an elaborate defense. Typically, the student was one who had done no work during the course and had simply tried to bluff on the final examination. Fred is so brilliant that he could transform almost any incoherent nonsense into a novel and insightful solution to a statistical problem. The rest of us were both impressed by Mosteller's genius in being able to make silk purses out of sows' ears and dismayed by the time and energy consumed in his trying to defend students who we had no doubt deserved to fail.

However, in at least one case, I know that Fred's tactics prevented a serious miscarriage of justice. A student, who was older and more mature than the others, had submitted an examination that the rest of us agreed was a failure. Fred, as usual, discovered a variety of reasons for considering the student's answers brilliant and innovative. Fred stuck to his position with such tenacity that the rest of us finally gave in and reluctantly agreed to Fred's demand that we change the grade from an F to an A. I later got to know this student—we collaborated on a publication—and realized that he had a very solid grasp of statistics. He also was very independent-minded and had very deep philosophical differences with many of the current statistical approaches that we had been teaching in the course. His deviant answers on our examination were due not to ignorance or lack of understanding, but rather to his strong convictions that the major statisticians were wrong.

Many students, especially psychology majors, have trouble trying to comprehend important statistical ideas. I remember that during one of my lectures on multiple regression some of the students expressed frustration at not understanding such things as the regression plane and regression coefficients. After the lecture, Fred took me up to his office, produced a wooden model of a regression surface, and suggested that I might want to bring it to my next lecture. I balked at this suggestion because I believed that the wooden model would not help. Fred, gently but firmly, insisted that I give it a try.

So at my next lecture, I dutifully brought in the wooden model and set it on a table in the front of the room. Without much conviction, I told the students that the model might help some of them better grasp the ideas behind multiple regression. A student who had been having trouble with regression excitedly jumped up and shouted, "So that's what it's all about!" Somehow this crude wooden model transformed this student's confusion into a confident grasp of the statistical model. From that moment on, this student had no trouble with the course. He is now a prominent psychologist.

In 1949, Fred played a key role in arranging a conference on "Mathematical Models in the Social Sciences," sponsored jointly by the Laboratory of

Social Relations and the Department of Mathematics. The main speakers included the distinguished mathematicians Norbert Wiener and John von Neumann, and Nicholas Rashevsky (a prominent mathematical biologist). Doris Entwisle recalls Fred's comments on how Wiener listened only to von Neumann and slept on stage when others spoke. The conference was remarkable as one of the early attempts to bring formal models into the social sciences, an activity in which Fred played a pivotal role over the next two decades.

Some Early Collaborative Research Projects

Among Fred's research projects during his early years in Social Relations was the pioneering empirical work on measuring utility functions (P24). We describe the technical details of this contribution in Chapter 5, but the following excerpt from the March 1949 issue of the Laboratory of Social Relations Bulletin suggests yet another facet of his activities:

> If it be granted that utility has a tendency to diminish as we acquire successive units of a good, it is evident that it is the importance attached to the possession of a single additional unit that determines what we are willing to give for something. This principle of valuation economists call by a name that sounds strange and wonderful, namely, "marginal utility." It means the utility derived from the possession of a single additional unit of a given stock of goods.
>
> Dr. Frederick Mosteller has set out here in our laboratory to bring the question of the relative value or utility of various amounts of money out of the literary realm of guess and talk and into the clearer domain of experimental science. His idea is to find out the value which various persons put on various additional amounts of money by determining experimentally how much of what they have they are willing to pay for long and short chances at winning the various amounts.
>
> The experiment itself consists in finding for each person the set of odds he wants in order to buy a chance on a certain piece of money for a nickel. The experimenter determines what chance of winning the subject has to have before he stops refusing to buy the chance on some bigger piece of money and starts taking.
>
> . . .
>
> Mosteller points out that similar experiments have been done in the past, where the interpretation was in terms of "misunderstanding" of the odds, rather than in terms of how a man views the value of the prize. He admits that "misunderstanding" of the odds may figure prominently in gambling behavior. In his

experiment, measures were taken to have the accurate odds before the subject, but still, superstitious behavior appeared, and showed that some non-rational factor must be at work.

Mosteller suggests that the truth of the matter must lie somewhere in between the two contentions: probably both misunderstanding and also the changing valuations placed on various amounts of money are important in establishing the shape of the utility curve.

In conclusion, it is pointed out that the experiment has so far shown that it is possible to measure utility in this fashion, it has shown that in some cases pretty good predictions can be made of specific behavior of an individual or a class, on the basis of the curves thus established. Still, the over-all shape of the marginal utility curve has not been established for any individual, or for people in general, because to establish such a thing, one would have to run the whole gamut of offers from zero to something well above ten thousand dollars, and the experimenters were not prepared to handle the cost of such an experiment.

Although the author of this description is not identified, some turns of phrase remind us of Fred's writing. Over the years, he has continued to stress to students and colleagues the importance of preparing careful write-ups of empirical work, as a step toward formal publication.

At about this time Robert Bush, who had been trained as a physicist, joined Social Relations as a postdoctoral fellow and soon thereafter entered into a major collaboration with Fred. In describing Bush's years at Harvard, Fred reveals much about their collaboration and Fred's own unique role in Social Relations (P97, pp. 165–167):

Although we do not know how [Bob] decided to come to Harvard's three-year-old Department and Laboratory of Social Relations, it was an exciting and stimulating place in 1949 teeming with clinical and social psychologists, sociologists, and anthropologists interacting vigorously and productively. The research ran from totally nonquantitative, nonempirical social philosophy through wide-open anthropological field studies, to the tightest laboratory experiments; from completely empirical studies, to totally theoretical mathematical ones; and from research in hospitals and in mental institutions to research in work camps and on psychodrama. Bob found it easy to join in seminars and courses with graduate students and in the discussions of the faculty. By participating in many of the informal student-faculty study groups, he learned a lot of psychology and social science and quickly made a reputation for himself that ultimately led to his appointment to the staff.

In those days, preliminary tiptoeing explorations for a special visitor included making sure that someone on the grounds would be responsible for making the visitor welcome and facilitating his work, and in some cases the possibility of working together was raised. As a recent Ph.D. in mathematics myself and a member of the Department of Social Relations with some ties to the Social Science Research Council, it was natural that these inquiries and suggestions would drift my way. . . .

As a postdoctoral fellow, Bob's primary project was to do a good deal of reading and to attend lectures and seminars, which he did with a will. On the research side, I suggested three possible areas with a view to our working together on one of them. They were (1) the study of problem solving in small groups (related to departmental work in progress by R. Freed Bales, earlier my office-mate), (2) finding relations between various psychological scaling methods through theory and experiment (based on a course I had given in psychometric methods), and (3) developing mathematical models for learning (based on some data on the relief [of pain] from successive doses of analgesics brought to my attention by Dr. Henry K. Beecher of Massachusetts General Hospital). As soon as Bob came, we spent a few days looking into these problems.

As for the first, we couldn't see how to get a sharp mathematical wedge into the small-group area .... Bob didn't care for the scaling problem even though it looked tractable both mathematically and experimentally—he said he wanted something more social. Again this is amusing because Stevens's later work was directed exactly to the social, or at least societal, uses of scaling. It all belongs to the *New Yorker*'s "Department of the Clouded Crystal Ball." As for the third area, we both saw ways to start on probabilistic models for learning. And so in a matter of three days, we chose and began to work on a problem that turned into years of effort. . . .

With his fellowship, Bob was free most any time, and my duties during his first few years at Harvard were only moderate. We were frequently able to spend half a day at a time together on research and then half a day apart. Memoranda were written very rapidly. Long phone calls were frequent. We had the aid of a well-trained psychologist and mathematician, Doris Entwisle, as mathematical assistant. By 1965, she was teaching in both the Departments of Social Relations and Electrical Engineering at The Johns Hopkins University. Bob pointed out to her as she launched on her own independent research career that it was going to be hard for her to get the kind of assistance we had had. . . .

> After Bob took a teaching post in the Department of Social Relations, his time was not quite so free, and because I was acting chairman in 1953–1954 we had a terrible time finishing our book before I went on sabbatical, but we managed, mainly by working very late at night. In writing the book, we usually worked out the theory and the statistical analysis together and then one of us would take the pieces and draft a chapter; the other would later revise it. ...

Doris Entwisle also recalls how Fred worked with students and colleagues, including Bob Bush, during this period:

> Fred's biggest gift to me was having high expectations. When I worked in his office, he treated me as a colleague. For that matter he respected anyone who deserved respect. He particularly respected students who tried. He was exceedingly patient with students and colleagues who came for help, no matter what the level of their questions. He also respected people who worked hard. I recall one observation he made about Bob Bush. He told me that Bob started every day with a laundry list of tasks, and that Bob would not stop work until the entire list was finished. In telling me about this trait of Bob's, he said: "Why, do you know that, on the night he was throwing a [particular] party, Bob was downstairs working on the last task on his daily list while his party was going on upstairs?"
>
> I recall in particular two graduate students whose names are, if not household words, at least pretty familiar. One was Bob Solow. He used to come by periodically to report on his progress in a reading course he was taking under Fred's direction. He would fill the office blackboard in short order, using just a few handwritten notes. He had the demeanor of an authority right from the start.
>
> Another student taking independent study was Tom Lehrer. He was brilliant in a different way, but also a spellbinder at the board. He was in conflict about choosing a career. When he eventually chose music and Broadway over applied math, I was not surprised.

[Robert Solow went on to a distinguished career as an economist at M.I.T. and received the Nobel Prize for Economics in 1987. Tom Lehrer returned to Harvard periodically over the next 16 years, and ultimately he became a professor at the University of California at Santa Cruz, as well as a celebrated song-writer and entertainer.]

Some Other Recollections

Ray Hyman recalls other stories from this period:

> Fred took me to lunch at a popular restaurant in Harvard Square. During the course of the lunch many faculty members stopped by to chat, and Fred introduced me to several of them. One of these was Ed Borgatta, a sociologist.
>
> Ed dropped a manuscript on the table in front of me and asked me to look at it and give him an opinion. While Ed, Fred, and other passersby were talking, I dutifully began making my way through Borgatta's manuscript, which bore the enigmatic title, "Sidesteps toward a nonspecial theory." The first few pages contained an introduction and footnotes that referred to previous attempts to come up with major theories of human behavior. Both in the text and in the elaborate footnotes Borgatta discussed Freud, Merton, Parsons, Sorokin, Adler, Horney, and others in ways suggesting that, while these pioneers had made important contributions, they had somehow missed the obvious and crucial feature of human nature. Of course, Borgatta's new theory was going to repair this defect.
>
> When the manuscript finally got to Borgatta's theory, I began to realize something was radically wrong. In the interval between accepting the position at Harvard and my arrival on the scene, I had already discovered that many of my new colleagues wrote and dealt with topics in abstract and obscure terminology that I found almost impossible to comprehend. So I was not surprised to find that I had not the slightest idea of what Borgatta was saying in the first several pages of the manuscript. But when he began describing his theory of "deumbilification," I considered it to be bizarre.
>
> I leaned over and whispered into Mosteller's ear, "Is this a joke or something? He can't intend this to be serious, can he?" Mosteller gave me a stern look. "Shh," he warned me. "Be careful. This guy has put all he has into this manuscript. He sees it as his major contribution to science. He is passionately attached to his theory, and, whatever you do, do not say anything to insult him or we will have a terrible scene on our hands."
>
> Soon after Fred's stern warning, Borgatta turned to me and asked for my opinion. While Fred kept a straight face, I spent the rest of an awkward lunch trying to be noncommittal and trying to find some complimentary things to say about some aspects of the paper.
>
> Only later did I learn that Borgatta was spoofing contemporary social science theorizing. The paper was published in *Psychological Review*, 1954. Many readers took it seriously and

wrote outraged rebuttals. It has become one of the classic spoofs of scientific literature.

Fred loved playing poker with a group of faculty colleagues. When I arrived, I was invited to join the group. Each participant chose the type of poker we would play on his deal. Usually the dealer designated one or more cards as wild. Fred was especially fond of making up versions that involved various combinations of high and low and several wild cards. I believe that his knowledge of the odds involving most of the standard versions of poker made them too dull and predictable for him. He seemed to enjoy variations in which it was impossible to anticipate the odds.

Soon after I had joined the Harvard faculty, Lou Guttman, the Israeli sociologist and psychometrician, spent a year in the Department of Social Relations as a Visiting Professor. During his visit, Lou participated in our weekly poker sessions. Lou claimed he knew little about poker, especially about the unorthodox variations that typified our play. I think he protested too much. I suspect that he pretended ignorance to gain an advantage over the rest of us. He walked away from most of our sessions as a winner.

On one occasion, Lou was the center of one of the few disputes that erupted over the outcome of a hand. The game was a variation of high-low. Lou had won the high hand, but had tied for the low hand. According to Mosteller, Lou had lost because the accepted rules were that one who had declared high and low had to unambiguously win both the high and low hands. Lou strongly protested that he should be the winner because no one had stated the rule before the hand had commenced.

The next day, the participants found in our mailboxes a manuscript that Fred had written to resolve how the stakes of the game should be divided given the misunderstanding about the rule in the high-low game. The manuscript used several different estimation procedures and assumptions to arrive at a set of converging conclusions. (Somewhere in my files I still have this manuscript, but I have been unable to locate it.)

Although I had performed magic professionally before I became a psychologist, I did not realize, until some months after I had met him, that Fred Mosteller was an amateur magician who had invented and contributed magical effects to magic journals. When I first discovered that he was a magician, I realized that this was the Mosteller of the famed "Mosteller Principle," a clever idea which I had used in some of my magical performances. Neither I, nor others, had much success in persuading Fred to demonstrate magic.

Fred continued to retain his interest in magic and many years later gave encouragement to another young magician, Persi Diaconis.

Participation in Henry Beecher's Pain Studies at the Massachusetts General Hospital

Academic anesthesia first learned about Fred Mosteller in the late 1940s when the late Henry K. Beecher, at the Massachusetts General Hospital (MGH), enlisted Fred's expertise to analyze data pertaining to subjective responses in humans when given a variety of drugs, particularly analgesics for postoperative pain. Thus began a career-long involvement in biomedical research. Later on, again with Beecher, Fred's analysis of postoperative mortality once more gave the first accurate picture of events, particularly the overall death rate and specifically the effects of curare.

Arthur Keats worked with Beecher and Mosteller on the pain studies and relates that "no matter what the problem was, he'd set up a theoretical model and come up with answer. All I wanted to know was whether the results were statistically significant, to put a P value on the study, but Fred never said just 'do a t test.' "

Louis Lasagna, now Dean of the Graduate School of Biomedical Sciences at Tufts University, arrived at the MGH after Fred had been on board for a while as a part-time consultant, and he wasn't sure at first what to make of Fred's advice:

> Harry [Beecher] explained that Fred was an exceedingly sophisticated and respected statistician, and that his soft-spoken and seemingly diffident manner belied a fundamental toughness of spirit. As in so many other matters, Harry was right on target.
>
> One of the problems we worked on was the contributions of discriminating and non-discriminating subjects. In those days, crossover designs were almost mandatory in Harry's view, so that it was possible to compare subjects' evaluations of the treatments which they received. Mosteller called to my attention the beautifully logical and persuasive stratagem originally proposed by McNemar. In this analysis, one in essence throws out the subjects who have responded equally to the two treatments being compared, regardless of whether both treatments were successful or failures. In other words, only subjects who responded differentially to the treatments were analyzed. In some studies, it turns out that only a handful of patients are in fact discriminators, but if one dilutes them out by lumping them together with non-discriminators, one ends up with statistically insignificant results. On the other hand, statistical significance can be achieved by not lumping them.

> Fred and I collaborated on the first paper in the literature to try to tease out psychological determinants of the placebo response. I picked out patients who consistently failed to get relief from placebos and contrasted them with patients who consistently got relief with the placebo. (Most patients were inconsistent, but I hoped to dramatize the differences by this contrast.) Sure enough, there were differences, but in retrospect it was a mistake to coin a term like "placebo reactor," because in fact people shouldn't be pigeon-holed as reactors or non-reactors. Rather, one should talk about "placebo-reactivity" which, like honesty or susceptibility to salesmen, is distributed along a continuum. Some people are more likely to be reactive than others, I believe, but the specific situation is also an important variable. (It also doesn't help that some "placebo responses" are simply the result of time changes, independent of suggestibility, attitude toward drugs, hospitals, health professionals, etc.)

Other members of Beecher's group included John Bunker and Henrik Bendixen who, with Leroy Vandam at Harvard, were later instrumental in getting Fred involved in the National Halothane Study (see the discussion in Section 7.3 below). As we noted above, Fred also credits this work with Beecher as a stimulus for his collaboration with Bob Bush on stochastic learning models.

The Early 1950s—Statistics and Computing

Cleo Youtz, Fred's longtime research assistant and collaborator, began her work with Fred in 1952. She recalls:

> Fred's office was well equipped by the standards of those days. He had a Royal manual typewriter, a Sundstrand (ten-key) electric adding machine, a ten-bank electric Monroe calculator, and good tables of the binomial, normal, gamma, beta, logarithms, powers of *e*, etc., and later Smirnov's tables of Student's t.
>
> The calculator was one of the more sophisticated ones—it had automatic multiplication and division. The routine was to enter the multiplicand on the keyboard, press the plus bar, clear the keyboard, enter the multiplier, pull the lever for multiplication, and wait for the wheels inside to spin and the carriage to shift and finally come to rest so that we could see the answer and write it down. For division one followed a reverse routine. The dividend and divisor had to be aligned to minimize the number of subtractions. If the machine jammed during an operation, the remedy was to detach the electric cord and turn the knob on the right until it felt in balance again. If this worked, we started

the calculation over; if not, a repair man was called—with a delay of possibly a day. Sometimes one of the wheels inside would slip a cog, giving an undetected incorrect answer. We devised various checks, almost as time-consuming as the original calculations. We could extract square roots—for large numbers, in a systematic way; for small numbers, Fred taught me to first make an estimate and with usually one or two iterations get the answer to the number of decimals that we wanted.

Working with this equipment was far from boring. We were always very enthusiastic about getting accurate numbers from our calculations. And we gloated when we found an error in one of the big tables. I recall finding a few in the National Bureau of Standards binomial table and normal table, and in Smirnov's t-tables.

Also, Fred was skilled with his slide rule. An editor once wrote to Fred for advice about the possibility of publishing tables of percentages that a research director in industry had put together. The director had found the tables useful for his own work and thought that others might also. The editor believed that answers could be had faster and with more accuracy from such tables than with a slide rule or calculating machine, and asked if Fred agreed. Fred gave it careful thought and replied:

> "I am not at all sure you can get percentages faster and more accurately from a table of percentage equivalents than from a slide rule or calculating machine. It depends partly on how often you use the same base. Also it depends on your business.
>
> "With a calculator that jammed three times, we got 6-decimal accuracy in 3 minutes 45 seconds for 15 divisions such as those you describe, including writing them down, while it took 4 minutes 46 seconds with a very broken down slide rule (no slide) to get these divisions to 3 decimals. I suggest you write to . . . and get his opinion as well as that of others on their staff as to the advantages.
>
> "I don't know whether it's really worthwhile, honestly. Very borderline, I think."

Almost all of the many calculations for the Bush-Mosteller learning book (B5) were made on the desk calculator. In addition to data from real experiments, a hypothetical animal which they called a "stat-rat" was carried through a sequence of trials using random numbers. This was an early example of simulations where several properties of a sequence might be studied, such as runs of events. Doris Entwisle carried out many of these calculations. Calculations for one large table were made by David Hays on the Mark I high-speed machine in the Harvard Computation Laboratory.

Several years would pass before the desk calculator was abandoned. But even in the early 1950s, Fred was thinking where high-speed computation would lead. He later played a leading role in its development at Harvard, especially time-shared computing.

Fred has almost always left computer programming to others. However, in 1953–1954, before programming became widespread, he and Solomon Weinstock worked out a set of programmed chess rules. Their computer was pencil and paper. They had a one-half-move version and a one-and-one-half-move version. They did win a game or two against another player. It was abandoned because, without a computer, the whole thing was too time-consuming and somewhat unreliable in the sense that all possible paths could not be explored.

7.3 In the Department of Statistics

Starting the Department

Much of Fred's time in the early 1950s was spent trying to get a department of statistics established at Harvard. His talk, "How Do You Establish a New Department at Harvard University?" at the 25th anniversary celebration dinner (1982) of the Department of Statistics explains:

> [Sam] Stouffer brought me to Harvard in the Department of Social Relations. [My wife] Virginia and I came in the fall of 1946. I never had a letter from Stouffer about my appointment. He didn't write letters.
>
> Statistics, as now, was taught everywhere: Crum and Frickey in Economics; Saunders MacLane in Mathematics; von Mises in Applied Mathematics; Harold Thomas in Engineering; Kelley, Rulon, Tiedeman, and others in the Graduate School of Education; Hugo Muench in Biostatistics 1948–1949 following the retirement of E.B. Wilson, who invented confidence limits.
>
> December 1949: Ahlfors, Thomas, Middleton, Birkhoff, Redheffer, and I (all of Harvard) discussed the idea and I proposed that a Department of Statistics should be established; the Mathematics Department agreed.
>
> April 29, 1950: I had a letter from Ahlfors—Paul Buck (Provost) declined, much too narrow a field for Harvard.
>
> Nothing happened in the years 1950–1953.

In 1953 the Ford Foundation gave self-study money in the behavioral sciences to five universities including Harvard. As Acting Chairman of the Department of Social Relations, I was deeply involved. The chairman of the whole self-study enterprise across the five universities was W. Allen Wallis. During World War II, Virginia was Allen's secretary. The self-study committee recommended that there be a Department of Statistics. The visiting committee of the self-study program also recommended a department.

The next year (1954–1955) I went on sabbatical to the University of Chicago to finish the book on mathematical learning theory and to work with Jimmie Savage, Allen Wallis, David Wallace, and William Kruskal, and there I met John Gilbert [who later came to Harvard and collaborated with Fred on many projects].

In May 1955, Dean McGeorge Bundy wrote inviting me to serve on the Committee on Applied Mathematics and Statistics under Dean J. Van Vleck, with several very distinguished colleagues: H.H. Aiken, G. Birkhoff, G. Carrier, S. Goldstein, E.C. Kemble, W. Leontieff, D.H. Menzel, W.E. Moffitt, G. Orcutt, H.A. Thomas, Jr. We began meeting on October 14, 1955. The Subcommitee on Statistics consisted of Frederick Mosteller, Chairman, G. Birkhoff, R.C. Minnick, G. Orcutt, and H.A. Thomas, Jr. The final report of the subcommittee was made January 27, 1956. The full committee met Wednesday, March 14, 1956. Van Vleck reported on May 18, 1956, that Dean Bundy favored the idea but was concerned about resources.

The issue came down to whether I would go chair statistics at Chicago or stay here. The matter dragged on all fall. But on December 7 Dean Bundy agreed to establish the department and appointed Orcutt (Economics), Birkhoff (Mathematics), and me as the organizing committee.

We made W.G. Cochran our choice for the first new tenured appointment. Meanwhile I trotted around to Mathematics, Economics, Social Relations, and the Commitee on Educational Policy to get support for the Department.

On February 12, 1957 (Lincoln's Birthday), at the Faculty of Arts and Sciences Faculty Meeting, Van Vleck moved the establishment of the Department, Lynn Loomis (Mathematics) seconded it, Robert White supported it for Social Relations, Seymour Harris of Economics spoke for it—he spoke at every faculty meeting—and E. Newman and Arthur Smithies raised questions. Bundy said initially only graduate degrees would be given. The total time on floor of faculty meeting was 14 minutes.

On April 14, Cochran agreed to come. Subsequently, John

Pratt came and Dempster (who was also invited) went to Bell Labs.

Howard Raiffa agreed on November 26 to come to the Harvard Business School with part time in Statistics.

To come back to the original question—How do you establish a statistics department at Harvard?—VERY SLOWLY.

Introductory Probability and Statistics Texts

As the work of the Department of Statistics got under way, several activities received special attention. Because of his interest in introductory statistics courses, Fred began, in 1955, to work with a group of the Commission on Mathematics of the College Entrance Examination Board. The purpose of one of the subcommittees was to get the teaching of probability and statistics into the high schools. The subcommittee produced, in 1957, the book *Introductory Probability and Statistical Inference for Secondary Schools: An Experimental Course* (B6). Within two years the book passed through the experimental stage, and a revised edition (B7) was published in 1959. It was translated into Spanish in order to spread probability and statistics into Latin America.

Out of the Commission on Mathematics group the triumvirate Mosteller–Rourke–Thomas coalesced. (George Thomas was Professor in the Department of Mathematics at MIT, and Robert Rourke was Head of the Mathematics Department at the Kent School for boys.) Together they wrote the book *Probability with Statistical Applications* (B10), with offspring *Probability: A First Course* (B13) and *Probability and Statistics: Official Textbook for Continental Classroom* (B12).

Statistics graduate students at Harvard during the early years of the department had many opportunities to see the problems and materials in these introductory texts. They found their way into the introductory courses, and Fred often tried out specific problems and solutions on graduate students.

In Chapter 6, Judith Tanur describes Fred's work on NBC's Continental Classroom series on Probability and Statistics, broadcast in 1961. During and following the TV series Fred received and answered some 200 letters from viewers. Over the years letters have arrived at Harvard from high-school students seeking help for their science fairs, and Fred has always answered with suggestions. He was especially happy when one student subsequently wrote with the news that his project had been judged first. Fred replied, "Your letter and paper have certainly brightened my day. I am delighted with your work, I find it most instructive, and I am happy you have had success with it ...."

More recently (in 1984) a high-school student wrote asking for a suggestion for a science fair project. The letter was addressed simply to Harvard University, but within a few days it reached Fred's desk for reply. It began: "Dear Harvard, ... I write to you for some suggestions for the Science

Fair. My project deals with statistics, ... This is a long term project so I am looking for a project that would take about a month to complete ...." Fred replied with a six-page description of a possible project.

Fred continued his collaboration with Rourke until Bob's death in 1981. Bob had been the stimulus for the 1964 booklet *Fifty Challenging Problems in Probability with Solutions* (B22) and their later joint efforts *Sturdy Statistics* (B33a) and *Beginning Statistics with Data Analysis* (with Stephen Fienberg) (B48), both labors of love. (Regrettably, Bob didn't live to see the completion of *Beginning Statistics.*) All of these materials were tested in the Harvard classroom. Fred has always been a firm believer in checking out pedagogical materials on the intended audience and then revising the work accordingly.

The *Federalist* Project

Fred always brings interest and excitement to work. He has an insatiable curiosity to learn and stimulates a similar curiosity in others. Fred's project with David Wallace on the authorship of the *Federalist* papers shows how Fred organizes group efforts and maximizes the enthusiasm and efforts of others.

The project started in 1959. Cleo Youtz recalls Fred asking her to organize the preparation of a concordance of *The Federalist* papers so that they could use it in an analysis to determine which of the "disputed" papers were written by Hamilton and which by Madison—both had claimed authorship. Fred knew it would take a long time before a concordance could be produced from a high-speed computer (programming was fairly new; FORTRAN arrived in 1957). In the meantime, how could they get word counts? Fortunately, at this time the Department had a couple of electric typewriters. Cleo helped to rig these with rolls of adding machine tape. Typists typed one word to a line, spaced vertically a few times, and typed the next word. Then they cut the tape into strips, one word to a strip. Sorting and counting came next. There was much excitement around the Department a few days later when the marker word *upon* emerged to help Fred and David distinguish Hamilton's writings from those of Madison.

While the typing and sorting progressed, Wayne Wiitanen, Harvey Wilson, and Robert Hoodes worked under the direction of Al Beaton on a program to produce a computerized concordance. Ideas from Fred and Dave came faster and faster. About 100 people worked on the project at one time or another. Fred and Dave planned the analyses, and Miles Davis handled the programming of the high-speed calculations. Ivor Francis's Ph.D. thesis grew out of his work on the project, and other statistics graduate students who contributed along the way included Roger Carlson, Robert Elashoff, Robert Kleyle, Charles Odoroff, and Sam Rao.

From the National Halothane Study to *Discrete Multivariate Analysis*

Toward the end of the 1950s, it appeared that a newly introduced halogenated anesthetic, halothane, might be causing fatal hepatic necrosis (liver failure). In December 1961 the National Academy of Sciences–National Research Council, through its Committee on Anesthesia, granted funds for a multi-institution study of the problem. Naturally the Committee turned to Fred to help in the design of this project and its subsequent statistical scrutiny. We use the word "naturally" because this involvement brought Fred together again with several anesthesiologists with whom he had worked in the Beecher pain studies.

One of the major statistical problems faced at the outset of the Halothane Study was whether to conduct a randomized clinical trial or an observational study. They started out with a randomized clinical trial, but shortly had to abandon it on ethical grounds, as John Bunker describes (1972, pp. 102–103):

> To resolve [whether halothane was linked with fatal liver failure] it was suggested that recognition of a link would require "a very large series of carefully matched surgical patients, the anesthetic (halothane as compared with one or a group of other drugs) chosen in a random fashion and the study planned in advance." [Bunker and Blumenfeld, 1963] Suddenly there were a dozen cases of fatal liver failure after halothane anesthesia, and another witch hunt was on. This time the anesthesiologists were ready. Halothane had proved itself clinically in several million cases, but the possibility of serious toxicity had been raised. It was clearly a public health problem sufficiently serious to justify a major clinical investigation, perhaps an ideal subject for a large scale randomized clinical trial. And so The National Halothane Study was born.
>
> The study was received with widespread enthusiasm by the medical and scientific community: by the anesthesiologists, who, after all, had a very deep concern for the future of their favorite anesthetic; by the pathologists, many of whom welcomed the opportunity of acquiring a large collection of morphological material for detailed study; and by the statisticians, who had just acquired the use of high-speed computers and were eager to apply them to a large scale epidemiological study of exactly the kind contemplated.
>
> The statisticians were particularly enthusiastic at the prospect of a randomized clinical trial in the operating room, where they felt this methodology was needed yet unproved. It was never completed, however, for just as the clinical trial began another case of fatal liver failure occurred and it was

> cancelled on ethical grounds. How could one justify the "experimental" administration of a drug—halothane—to patients in view of its imputed risk? Ironically, the widespread clinical use of halothane continued while the Halothane Committee searched for other ways to answer the question of whether the use of halothane was linked with fatal liver failure.
>
> With the prospective randomized clinical trial precluded, a retrospective study of past data was reluctantly selected as the only alternative. It seemed likely that the effect under study—fatal postoperative liver failure—might be very rare, possibly one in 10,000 administrations of anesthetics, and the nonrandom selection of patients for anesthesia and surgery might well defeat the study. One of the first orders of business for the statisticians, therefore, was to try to determine how anesthesiologists select anesthetic agents for a particular patient. The statisticians pursued this goal with vigor and determination, and we as anesthesiologists tried to answer all their questions.

Fred was not only a participant from the beginning, he was centrally involved in the whole project from the beginning through to the end—and beyond. He and Lincoln Moses were the only non-M.D.s on the Committee itself, but they were joined in the analytical work by a distinguished group of statisticians that included Bill Brown, John Gilbert, and John Tukey. The statistical work lay at the heart of the project, and graduate students were recruited to help in the analysis. These included Tukey's student Morven Gentleman, and Fred's student Yvonne Bishop.

Lincoln Moses recalls a letter Fred wrote to another key figure during the planning phase of the study:

> The letter's message was that Fred was alarmed, and the addressee should be alarmed too. For years I kept a copy of it, as a model of how to inform a client that he was approaching very thin ice, and how deadly that would be; it was a most eloquently alarming missive from the typically unflappable Fred. A model indeed! I recall an urgent phone call from him to me when the analyses were beginning to show signs of becoming exotic; he said something to the effect of "make a two-way table, showing institutions as rows and anesthetic agents as columns and see what it says!"

Yvonne Bishop went to work full-time on the project and worked with Fred on what became several chapters in the Halothane Study report (B24) and the basis of her doctoral dissertation on the analysis of multi-way cross-classified categorical data using loglinear models. Her effort brought together a computational approach to contingency tables suggested by Deming and Stephan (1940) with the structure of loglinear models suggested by Birch (1963). The key was the use of the computer to analyze large multi-way arrays, a task that previously would have been considered unthinkable. Yvonne did much of this work at the Center for Advanced Study in the Behavioral Sciences (CASBS), near Stanford University, and had access to the Stanford computer through John Gilbert, the CASBS's resident statistician.

Fred was especially impressed by John Gilbert's work and made arrangements for him to move to Harvard in a special staff position. Yvonne Bishop also returned to Harvard to work on her dissertation. At this stage the statistical analysis job on the Halothane Study was not yet complete. Lincoln Moses remembers receiving a telephone call one day from Fred:

> Fred asked me, "Who is the staff now?" My answer was vague—we were all working on it in our spare time. Fred was alarmed! He immediately set up solid computing arrangements at Harvard. The study moved swiftly to completion. That episode informed me of a most valuable question, one that I have occasionally used quite helpfully since: "Who is the staff?"

Fred succeeded in getting several others around Harvard to work on problems related to the Halothane Study. Stephen Fienberg recalls Fred asking him to work on "a little problem" on smoothing counts in 1966. The little problem was yet another approach to the major methodological issue in the Halothane Study, and it led to Fienberg's collaboration with John Gilbert on the geometry of 2×2 tables and with Paul Holland, then an assistant professor in the Department of Statistics, on Bayesian estimation in contingency tables. These materials, along with another collaborative piece with Yvonne Bishop on triangular tables, all found their way into Fienberg's Ph.D. dissertation.

When the Halothane Study report was completed in 1969, Fred appeared as author or coauthor of several chapters in Part IV, "The Study of Death Rates": Chapters 1, 3, and 8 and part of Chapter 2. In addition, he was one of the four editors (with Bunker, Forrest, and Vandam) for the entire study volume, and his participation in the whole project was continuous, searching, and diverse.

The surprising finding in the study was that halothane, "which was suspect at the beginning of the study, emerged as a definitely safe and probably superior anesthetic agent." (B27, B28, B30) The study also revealed differences from hospital to hospital in postoperative death rates, which can only be partially explained by the differences in the patient populations. Later studies confirmed the results of the Halothane Study and added further information on the sizes of the differences and on the kinds of procedures for which the variation is greatest.

Fred was President of the American Statistical Association in 1967, and he decided to use his presidential address at the annual meeting (see P73) as a vehicle to describe all of the different ongoing contingency table research projects in the Department of Statistics. This was both a way to stimulate work in what he thought was an important new area of statistical research, and also an opportunity to highlight the work of various students.

In the spring of 1968, following the publication of his ASA presidential address and as the National Halothane Study neared completion, Fred gathered together a group of graduate students and junior faculty one evening at his home, in order to discuss the preparation of a book that would pull together all of the research that had been done in the Department over the preceding years on the analysis of contingency tables. They drew up a list of problems that still needed to be examined, and over the next several years various members of the group wrote papers and drafted chapters for what ultimately was published as *Discrete Multivariate Analysis: Theory and Practice* (B39) in 1975. In the end Yvonne Bishop, Stephen Fienberg, and Paul Holland became the principal authors, and Richard Light (another Mosteller Ph.D. student) helped with the chapter on measures of association and agreement. But Fred's hand can be seen in virtually every chapter and virtually every example. His suggestions and skilled editing were responsible for most of the book's special features, such as the extensive use of substantive examples and the index of data sets at the end of the book, one of the first of its kind.

Few parts of the final book resembled the research reports and other materials with which the group began in 1968. The Halothane Study and the methodology developed for it spawned considerable interest in the statistical community. In particular, Leo Goodman wrote an important series of papers building on results in Yvonne Bishop's thesis, and he and Stephen Fienberg supervised the dissertation work of Shelby Haberman at the University of Chicago, which contained the first systematic treatment of log-linear model theory from a coordinate-free perspective. Throughout the preparation of "the jolly green giant," as the book is affectionately called by its authors and others (because of the bold green cover and its size), Fred did much to hold the project together, arranging for occasional visits by Fienberg to the Department and offering encouragement and support in other ways. Ultimately the book blended this new literature with the earlier approaches, including several previously unpublished methods and applications.

Mosteller and Tukey on Data Analysis

Fred and John Tukey have collaborated on a number of research projects, beginning with John's assistance to Fred on his Ph.D. dissertation. Most of those who were around the Department of Statistics at Harvard during the 1960s will recall the famous set of loose-leaf notebooks that sat on the bookshelf behind Fred's office desk and contained working drafts of material on statistical methods and data analysis topics. Many of Fred's research assistants worked on different pieces of this project.

Stephen Fienberg recalls his first meeting as a graduate student with Fred in September 1964:

> When I arrived in Cambridge, I reported to the Department, which was located up on the third floor of 2 Divinity Avenue, and, since Fred had written to me that I was to be one of his research assistants, I stopped by to meet him. He was busy in a meeting but asked if I could join him for lunch. At noon we walked over to the Faculty Club, and, as we sat down at a table, Fred pulled an envelope out of his jacket pocket, with some formulas scrawled on the back. "Tukey sent this to me last week," he said. "We need to work out the details of his idea." (At the time I didn't know who this guy Tukey was!) Fred then spent the rest of lunch explaining to me what the problem was all about. This was my introduction to the assessment of probability assessors.
>
> For the next several weeks I struggled with little or no success to translate the material from the back of Tukey's envelope into the framework Fred had described. Fred would have me write up notes on what we had done to date, and then I was supposed to go away and work on the problem some more. I was getting discouraged, but Fred insisted that we organize our notes into a technical report, and I was assigned to prepare the first draft. Many months later, after I had made the link to an obscure little paper by de Finetti in a volume edited by I. J. Good, we completed the technical report. A copy went into the files of the Mosteller-Tukey project on Fred's shelf. We never really talked about the topic again, and, as far as I have been able to discern, it never appeared in any of Fred and John's joint publications. I viewed the enterprise as a failure and thought myself fortunate when Fred asked me to work on a different problem, also drawn from the project with Tukey.
>
> About 15 years later, I attended a session at a conference in which Morris DeGroot discussed someone's paper dealing with the quality of probability forecasts. As I listened to his comments, I suddenly recalled the key formula that came off the back of Tukey's envelope, and I joined in the discussion. I thought: "Fred would be pleased to learn that I had finally understood what John had been trying to do. Maybe that was why he made me work so hard to write up the results." With the 1965 technical report as a starting point, DeGroot and I went on to tackle the problem from first principles, and we soon found ourselves working on a series of exciting papers. Coming back to a problem after many years and working on it from a somewhat different perspective was also something I

had learned from Fred.

Around this time, in 1965, Fred arranged with Martin Schatzoff of IBM, who had received his Ph.D. from the Statistics Department, for instruction in FORTRAN programming with hands-on experience on IBM consoles. Fred brought Cleo Youtz, along with his wife Virginia and their son Bill. This was the beginning of Bill's successful career in computing, and he helped out on the project with Tukey in a number of ways. Although Fred continued to have most computing tasks done by others, he now had a much clearer sense of what was possible and how it should be done. Bill Mosteller recalls:

> Although Dad took the Fortran class, he never actually wrote any further Fortran programs that I know of. He did teach me a concept about computers which has stood me in good stead ever since. "*Distrust any result produced by a computer.*" I learned to manually calculate results produced by my programs and compare them. I learned to print intermediate results, in defense, to simplify determining where the calculation went off track. I hated it, but my programs produce much more accurate results. I learned not to announce conclusions from the first run that gets completed, and have never regretted that lesson.
>
> The testing approach I use today is based on that outlook. I generally assume that any program I write does not work, and test accordingly. I believe that my work is better for that attitude. Fortunately, his insistence on accuracy in computer programs was coupled with a willingness to invest the time necessary for that accuracy. The time pressure so often associated with computing work was missing. He was always willing to spend the time necessary to make sure we both understood what was expected and what was being prepared, and this made him easy to work with.

Most of those who were graduate students in the mid-1960s were never really sure what the entire data-analysis project encompassed, although many helped out with different pieces. For example, Steven Fosburg worked with Fred on a computer-based reanalysis of data from the *Federalist* papers using jackknife methods. He also worked some on a section on Bayesian methods and conjugate priors. Major parts of the project were organized for a handbook chapter, a draft of which was circulated to faculty and students in the Department for comments and revised yet another time.

In his conversation with Tukey (P168), Fred describes this project in his typical understated way (p. 143):

> Gardner Lindzey had asked John and me to write a statistical chapter (Mosteller and Tukey, 1968 [P76]) for the second edition

> of the *Handbook of Social Psychology.* John and I worked on that very hard and we wrote much too much. Whereupon Lindzey took a share of it and we were left then with a considerable extra bundle. So we decided to put the bundle together and create a book around it called *Data Analysis and Regression.* And it does include some information about robust methods as well as more classical kinds of techniques. It especially has a substantial discussion of regression and the difficulties and hazards associated with multiple regression. Essentially it says a lot about how little you can do with regression as well as how much. That's an important feature of the book.

The book also drew on ideas and problems that emanated from the Halothane Study. John Tukey (in the same conversation with Fred) describes that part of "the green book" as follows (pp. 142–143):

> And in that there is material on adjusting for broad categories (pages 240–257). The thing that people forget when they say, "Well, I'm going to cure the fact that there is this background variable. I'll dichotomize it, and then I'll look at the two halves of the dichotomy and sort of take what effect there seems to be in each half and pool them together." Forgetting that if the ratios of the fractions are different, if you sort of cut a distribution across under a knife, the centers of gravity are not going to be in the same place when the distribution is over here and cut there, as when it's over there and cut here. Dichotomizing helps, but it's not the whole answer.
>
> Being sure you make this correction—or any correction—accurately is not likely in human affairs, but you are a lot better off to make the correction than not. Correction for broad categories is one of the things that really did get into the green book, and I think it's right to direct people's attention to it, because it's a pretty widespread problem that people often sweep under the rug, saying: "Well, we'll divide it. We'll at least separate the people with low blood pressure from the people with not-low blood pressure and then not worry about the details anymore." We learned something about broad categories in the halothane study because the data was all collected on the 800,000 cases before one could get a hard look at it, and the ages had been coded in ten-year blocks. It turned out that the risk of death from surgery about doubled every ten years, so the distinction between sixty-one and sixty-nine, or seventy-two and seventy-nine, was a really important distinction. If we'd only had better age data, we would have been able to squeeze things a little more. And if we knew more about broad categories, we could have done a little better. I don't think it would have affected

> the overall conclusions, but it would have been nice to come nearer to getting out of the data what was in there waiting for us.

Fred worked on a number of other data analysis projects during this period, not all of which were directly linked to the project with Tukey. One morning Fred came into the office and discussed with Cleo Youtz an idea for work on the Goldbach conjecture—that every even number above four can be expressed as the sum of at least one pair of primes. Fred wanted to know how many different ways each even number could be so expressed—the Goldbach counts. He commissioned his son Bill to write a computer program to calculate the Goldbach counts. Cleo recalls:

> When Fred saw the first printout, he noticed right away that every third number had a large count; for example,

Even number	Goldbach counts
446	12
448	13
450	27
452	12
454	12
456	24
458	9

> I hurried to find my old number theory book to recall that primes greater than 3 have the form $6k + 1$ or $6k - 1$, k an integer. With this information, Fred produced a good intuitive argument for a large count being associated with every third number (P90). Later we found out that G.H. Hardy had worked on this problem.

Leaving the Chairmanship of the Department of Statistics

Fred served as Chairman of the Department of Statistics from its creation in 1957 until 1969. Even though he was constantly busy with several projects and outside activities, he still devoted a substantial amount of time to departmental activities.

Fred was the official thesis supervisor for a number of Ph.D. students in the Department of Statistics: Robert M. Elashoff, Joseph I. Naus, Joel Owen, Janet D. Elashoff, Ivor S. Francis, Mikiso Mizuki, Yvonne M. Bishop, Ralph B. D'Agostino, William B. Fairley, Stephen E. Fienberg, Gudmund R. Iversen, Richard J. Light, Farhad Mehran, Sanford Weisberg, Richard W. Hill, Michael A. Stoto, Judith L.F. Strenio, Judith D. Singer, and Katherine Godfrey. Fred worked with most of these students during the period that he was Chairman. In addition, Fred was an unofficial supervisor for several others, and he gave extensive advice to many graduate students inside and outside the Department.

In the late 1960s the Department had outgrown its quarters in the top floor of the stack space for the Harvard-Yenching Library at 2 Divinity Avenue, and it set up a satellite operation in Palfrey House, a block to the north, which housed most of the graduate students and a few faculty and visitors. During this period the program of graduate courses began to include a special focus on data analysis. Yet many of the graduate student traditions remained, including the practice of having one or two advanced graduate students take responsibility for organizing the weekly departmental colloquium. Fred was not simply delegating the work associated with arranging for speakers and serving as host during their visits. Rather, he viewed this activity as a way to help launch students as professionals. Several of those who were colloquium chairs as graduate students later became department chairs at other universities across the United States.

Around the time that Art Dempster succeeded him as chairman of the Department in 1969, Fred prepared a document summarizing various things he thought a graduate student should know something about but which were never part of any course of study. Some of this advice was specific to Harvard, as it was then, but most of it is as useful today, both at Harvard and elsewhere. We include major excerpts from this manuscript in an appendix to this chapter. Fred later rewrote part of this material for *The American Statistician* (P122), but we think the original captures Fred at his best in providing a model for graduate students to follow.

7.4 In the Kennedy School of Government and the Law School

The School of Government at Harvard began in 1936 as the Littauer School of Public Administration and became the John F. Kennedy School of Government after the assassination of President Kennedy. In 1968 Dean Don Price invited a few distinguished professors from other Harvard departments to help create an exciting new enterprise. Participants included Richard Neustadt, Thomas Schelling, Howard Raiffa, and Fred Mosteller. They proposed a school with a program dramatically different from the traditional 'public administration' model, one with a heavy emphasis on policy analysis, including statistics.

This small group of eminent senior statesmen met vigorously for two years, creating a new curriculum for a new degree program called MPP—Masters in Public Policy. Because of the small class size in early years (the first class, admitted in 1969, had only 25 students), faculty were able to experiment with new methods of teaching, including the case method.

Fred, predictably, led the effort to introduce new ideas into the statistics component of this program. Working with several younger colleagues, especially Will Fairley, who had written his Ph.D. dissertation under Fred's supervision a few years earlier, he created a statistics curriculum noticeably different from the traditional material taught in statistics departments. Fred's attitude was that most basic courses in statistics fell into one of two categories. Some courses, generally taught in statistics departments, were serious efforts to acquaint students with statistics as a research tool—how to do it. Other courses were "once over lightly" for students who had no intention of ever doing real research, who planned on careers as public managers, and who had little interest in statistics. These courses were often interesting, even entertaining, but Fred felt that no one should have any illusions that they prepared students to carry out, indeed even to understand, serious statistical work.

Fred began to create a brand new enterprise—a hybrid of these two activities. Along the way, he helped to stimulate development of new course materials, including case studies, and two books as well. One book, *Statistics and Public Policy* (edited with Will Fairley; Addison-Wesley, 1977) (B41), offered a series of illustrations of how statistical analysis helped to resolve policy dilemmas. It has been used regularly ever since in the Public Policy Program. A second book, *Data for Decisions* (co-authored with David Hoaglin, Richard Light, Bucknam McPeek, and Michael Stoto, 1982) (B44), laid out for public managers a series of tools for gathering information to help guide policy decisions. This second book is now widely used in executive programs at the Kennedy School and is popular among the decision makers who have passed through the School.

With Will Fairley, Fred created a syllabus for a full-year course in the Kennedy School and a shortened version for the Law School that "uncovered" (not covered!) the principal ideas and methods of statistics in collecting, exploring, and analyzing data. Although computation and a certain amount of mathematics were an essential part of these courses, students got the benefit of Fred's steady realism and deftness in applying statistics to live problems. A good example was Fred's classroom analysis of data supplied by Hans Zeisel on the celebrated trial of Dr. Benjamin Spock. Zeisel contended that Judge "G" presiding at the trial discriminated against women as jurors, and he challenged the all-male composition of his jury. The class looked at the proportion of women on the jury panels of a group of judges that included Judge "G." A binomial model for sex of juror selection for the panel and a normal-theory test of the hypothesis that Judge G's probability for selecting was the same as the others gave a z value of -8 standard deviations for Judge G's proportion. But did the binomial hold? In fact, Fred noted (paraphrased):

> Chi-square calculations of goodness of fit support the binomial assumption. We can conclude that the event (Judge G's proportion) is rare. The key is believing that we have the right

> measure of variation. This is vital. Only it allows us to be awed by −8 standard deviations.

(This example came from Hans Zeisel's chapter in *Statistics: A Guide to the Unknown* (B27) and was analyzed in detail in a chapter by Stephen Fienberg in *Statistics by Example: Detecting Patterns* (B36).)

Along similar lines, in another class Fred had the cautionary advice:

> Anyone giving very precise probability answers to a data analysis hasn't considered all sources of variation. He's got a pure model and he is deducing implications of it. In fact, its assumptions may not be met, even approximately.

Fred stressed for his audiences of future government policymakers and lawyers that they would *use* research and the statistical tools that interpret research. He took every opportunity to point out big public issues that consumed endless and fractious attention without benefit of research. The indictment he painted of a society that paid little attention to getting the facts about what was happening, on what worked, and on what didn't work has sown many seeds that have come to fruition in the work of graduates of these courses.

Fred's advice was often to get help from a real expert. For example, on sample surveys and with a characteristic down-home comment, he would say: "You'll probably want to buy one. The 'boy's own' survey is likely to be inferior." He stressed the importance of the standard tools of statistics in a classroom handout, "What is Statistics All About?":

> The most commonly used methods of statistical analysis are what we might call the standard models. Indeed, until the expression "models" became popular, they weren't known as models at all. They were just methods. But it is appropriate to understand that one of the principal offerings of statistics is a number of flexible all-purpose models which can be laid onto many a problem without much more than perfunctory consideration. Part of the idea of any field that pretends to help with research is that it supplies general-purpose methods. Special methods are usually very costly to create, and consequently we usually require some strong reason for doing them. Examples of standard all-purpose models are the theory of the sums of possibly correlated random variables, regression theory, analysis of variance, non-parametric methods, chi-square methods for contingency tables, maximum-likelihood estimation, or standard tests of significance. The standard families of probability distributions form another important set of models—the gamma distributions, the chi-square distributions, the beta distributions, the F-distributions, the t-distributions, and among the

> discrete ones, the binomial, the Poisson, the hypergeometric, the negative binomial, and the geometric distributions.
>
> These standard pieces of equipment give one a considerable armory for attacking new problems in a rather systematic way, for they all have the advantage, and it is an advantage, of providing a standard method of attack and providing standard answers to standard questions. So one can get started by reviewing with himself what questions among these standard ones he wants answered, if any, or if he needs something special, how does it differ from the more standard ones? This is a far, far better condition to be in than the man faced with a large body of data with no ideas how to get started at all. It is pitiful indeed to see people in this state. One finds them inventing the most astonishing statistics because they have nothing better to do. Much is gained by standard methods of organizing data.

Although standard is good, Fred also sought to understand the particular needs of the professional areas of policy analysis and law to which statistics would be applied. In these fields he noticed that people often had to do the best they could with secondary data and published statistics because time did not always permit them (though planning might have helped!) to carry out original data collection and analysis. This led to his attention to reasonable ways of assessing orders of magnitude for "unknown numbers" that were needed in a policy analysis or dispute. He published this contribution as a chapter in *Statistics and Public Policy* (B41, pp. 163–184).

A second example of Fred's interest in innovating for his "market"—as a good entrepreneur will do—was the staging of a mock "Adversary Hearing" by students on the topic of whether experimental results on weather modification justified a large-scale public program. The idea was that a responsible government administrator might be helped in reaching a decision by letting frankly opposed experts argue it out in a loosely formalized setting (P92).

Fred saw the arguments in favor of adversary settings in fact-finding (the decision maker has greater confidence that all information relevant to an issue will be brought out, and that flaws will be brought to his attention). In characteristic empirical fashion, he sought to test these arguments in a novel format. We see that Fred's interests lie, not only in the standard and the innovative tools of statistics, but also in the actual process by which the tools will be used and useful in policy and law.

Fred's lasting contributions to the Kennedy School go far beyond his particular teaching skills during any given year. He was one of very few people, *perhaps the only person*, who could make the case for including a serious study in statistics in a professional school of government at Harvard. And he was flexible and innovative in changing dramatically what most colleagues thought of, at that time, as "basic statistics." To this day, when the curriculum is reviewed, as it is regularly, senior professors fondly recall Fred's contributions. Indeed, there is still a friendly debate as to whether Fred's main contribution was *curriculum* development or *faculty* development at the Kennedy School, since he did so much of both simultaneously.

7.5 In the School of Public Health

Moving Across the River

Although Fred has had an interest in statistical applications to medicine and public health at least since his first interactions with Beecher, only in the 1970s did this interest manifest itself in formal relationships. Perhaps the first of these was the interdisciplinary Faculty Seminar on Health and Medicine, which Fred organized in 1973 with Howard Hiatt, who had recently become Dean of the School of Public Health.

Twice a month from October through May the Seminar met in the evening for a buffet dinner and a lecture, and then broke up into about a dozen on-going working groups on particular topics. Although known as the Faculty Seminar, it attracted a broad range of participants during its three academic years: faculty, graduate students, and academic visitors from many Harvard schools and departments, and researchers from other institutions in the Boston area. Fred helped to organize the working group on surgery, which usually held additional meetings in the weeks when the Seminar was not meeting and through the summer. John Bunker, who returned to Harvard that fall for a two-year visiting appointment in the School of Public Health, recalls:

> I naturally gravitated toward the working group on surgery. At the first meeting, almost before anything else had transpired, Fred stated that "of course, we will write a book"; and so it was that we embarked on *Costs, Risks, and Benefits of Surgery* (B40), which Oxford Press published for us four years later, and which I believe is considered a landmark in the new era of cost-effective medical decision making. Fred contributed as a co-author of three chapters and a co-editor of the volume itself.

The Seminar also produced a number of other publications from the other working groups. Among these were Anderson et al. (1980) and Neutra et al. (1978).

A second and closely-related activity involving Fred at the School of Public Health was the establishment of the Center for Analysis of Health Practices (CAHP), which was later incorporated into the Institute for Health Research. The Center was intended to revitalize the School's approach to policy issues related to medicine and public health and included staff members from a variety of disciplines from around Harvard. John Bunker played an important role with Fred in getting CAHP started, and CAHP was the formal sponsor of the Faculty Seminar. In 1975, Howard Frazier became Director of CAHP, and among the statisticians with appointments in CAHP for the 1975–1976 academic year were Art Dempster, Stephen Fienberg (visiting from the University of Minnesota), and Nan Laird.

During the 1975–1976 academic year, the Department of Biostatistics in the School of Public Health continued to search for a new chair, and various suggestions were made about formal links between the Department of Statistics, located on the main Harvard campus in Cambridge, and the Department of Biostatistics, located several miles away in Boston on the medical campus. Before long Fred had organized a committee to explore the possibility of a university-wide statistics department that not only would formally join Biostatistics and Statistics, but also would bring in the statisticians from the School of Business, the School of Education, the Kennedy School, and the School of Medicine. Fred tried his persuasive powers on his colleagues, but the decentralized nature of Harvard and the University's financial watchword, "Every tub on its own bottom," thwarted his best efforts. Most of the Deans involved and some of the statisticians were reluctant to create a department that had such a varied portfolio and reported to multiple Deans.

Paul Meier, who was visiting Harvard from the University of Chicago at the time, was a frequent visitor to the School of Public Health, and he predicted to Nan Laird that Fred would soon take over as Chair of Biostatistics. Nan assured Paul that this would never happen; in February 1977, Fred became Chair and held the position until 1981. Somehow Fred was convinced that helping to rebuild Biostatistics was important for Harvard and that he could make an important contribution by taking charge.

Fred retained his office in the Department of Statistics in Cambridge, but for a large chunk of his activities his appointment as Chair of Biostatistics signalled a formal move across the Charles River to the Medical Area of the University.

In the Department of Biostatistics

As soon as he agreed to assume the chair of Biostatistics early in 1977, Fred immediately set out to consolidate statistical activities at the School into the Department of Biostatistics and to create an academic community for biostatisticians and others engaged in quantitative activities at the School. One of his first priorities was to obtain more space and relocate "stray" faculty members within the Department. He initiated regular Department faculty meetings and began an active faculty recruitment program. Larry Thibodeau and Christine Waternaux joined the Department in 1977, Thomas Louis and James Ware in 1979.

In a concurrent development, the Sidney Farber Cancer Center (now the Dana-Farber Cancer Institute), one of the teaching hospitals of the Harvard Medical School, had nearly completed a search for a senior biostatistician.

The search committee had recommended Marvin Zelen, then at the State University of New York at Buffalo, for the position and proposed a joint appointment at the Center and the School of Public Health, but a snag developed. At Buffalo Zelen had founded a very active Statistical Laboratory, which was responsible for many long-term national clinical trials in cancer. It seemed likely that Zelen's move to Harvard would result in the dissolution of the Statistical Laboratory and disrupt the scientific and administrative support for these studies. Zelen proposed that the faculty of the Statistical Laboratory be evaluated by Harvard and that those who were both qualified and interested be given joint appointments in the School of Public Health and the Sidney Farber Cancer Center.

This unusual proposal received Fred's complete support, and he devoted his energies to persuading others in the School that it represented an unusual opportunity to strengthen the Department of Biostatistics. Fred discussed the proposal with the Biostatistics Department and prompted the remark by Bob Reed, "One is used to thinking of a whole baseball team moving with their coach, but a biostatistics team is a new idea." Fred's efforts turned out to be successful, and Zelen arrived for the 1977–1978 academic year with a "team" of nine new faculty, including Colin Begg, William Costello, Richard Gelber, James Hanley, Phil Lavin, John MacIntyre, William Mietlowski, David Schoenfeld and Ken Stanley. In 1978 two other Buffalo faculty, Stephen Lagakos and Marcello Pagano, made the move to Boston.

From the beginning of his chairmanship, Fred gave priority to the Biostatistics teaching program. He formed an ad hoc committee to revise the format and curriculum of Biostatistics 201 (required of all students at the School) and to introduce computing into the course. When Fred came to the Department, the only formal requirements for advanced degrees in biostatistics were those set out by the School. Fred formed another ad hoc committee (which subsequently became the Degree Program Committee) to formalize requirements for the various degree programs. The Curriculum Committee was formed shortly after the arrival of the Buffalo group and provided the structure for building a strong doctoral program in Biostatistics.

Another priority with Fred was to enlarge the participation of the Department in the scientific research activities of the medical area. In 1978 he set up the Biostatistics Consulting Service under the directorship of Yvonne Bishop. Subsequent directors have included Thomas Louis and Michael Feldstein. The Service (later renamed the Laboratory) has collaborated on short- and long-term projects ranging from prevention of myocardial infarction to broad-based health policy issues. It eventually was responsible for establishing statistical units at McLean Hospital and at Brigham and Women's Hospital. It served as an open-door statistical consulting activity and became an important forum for training biostatistics graduate students in the art and science of consulting. This training became so successful that the Department made coursework in consulting a requirement for all degree candidates.

Much of Fred's impact on biostatistics at the School of Public Health has resulted from relationships with junior faculty established through projects such as the *New England Journal of Medicine* (NEJM) Project, described below. Working with Fred on one of these projects was an education in the management of scientific research. Jim Ware recalls that in 1979, on joining the group writing the book *Biostatistics in Clinical Medicine* (B49), he was introduced to approaches and attitudes about joint research that made a lasting impression. Through a steady stream of memos, meetings, and informal conversations about progress, Fred conveyed a sense of purpose and attention to both concept and detail that was critical to the success of the writing effort.

Fred's administration of the Department was marked by frequent consultations with the faculty on new programs and directions. Although routine in Arts and Sciences departments, this approach is not typical in the Harvard Medical Area. Department chairs or heads wield great authority, and it is unusual for them to consult closely with members of their departments. However, Fred set in motion a series of senior faculty meetings to plan for the Department's future. One recommendation was that the Department should make a senior appointment in health policy. This recommendation led, in 1981, to the appointment of Milton Weinstein, who had previously served as a member of CAHP and on the faculty at the Kennedy School.

One of Fred's notable characteristics is his great loyalty and devotion to Harvard University. Fred stepped down from the chairmanship of Biostatistics in 1981 to assume the chairmanship of the Department of Health Policy and Management in the School. This department, one of the most important units in the School, was facing many difficult academic, financial, and administrative issues. Dean Howard Hiatt required help and Fred, as always, agreed to lend a hand. Fred once said that "No" was not a bad answer, just the second best answer to a request.

Fred left the Department of Biostatistics with an enhanced teaching program, a greatly enlarged and active graduate student population, and a thriving research program. In four years he led a "quiet revolution." Fred provided a great role model. He inspired many junior (and not so junior) faculty. He created innovative academic and research programs throughout the School.

The New England Journal of Medicine Project

Fred Mosteller has an unusual ability to organize other people and lead them together to some common goal. Indeed, veteran Mosteller-watchers might agree that this organizational talent is the most critical element in the way Fred's great personal abilities have been multiplied and extended in many group settings.

Fred's remarkable ability to exercise leadership on a grand scale without losing touch at the most detailed practical level was evident in the New England Journal Project. In 1977, Dr. Arnold Relman, editor of the *New England Journal of Medicine* (NEJM), suggested that a study of recent published papers could be used to educate his readers about which statistical methods were in greatest use, and to provide a clearer view of the role and importance of quantitative methods in biomedical research. Fred seized this idea with enthusiasm and began to elaborate a major research project. He soon had most of the Department of Biostatistics, both faculty and students, developing a brief abstract form and tabulating papers published in the Journal. Larry Thibodeau became statistical associate editor of the NEJM. All of this work served as the basis for a successful request for funding by the Rockefeller Foundation. Even before those funds arrived, the abstract form was substantially revised and extended, plans for detailed review of all material published in the Journal during 1978 and 1979 were in place, and a working group was meeting regularly to go over the initial findings, block out a series of papers and book chapters, and get ready to write up the results. In the end the project made only modest use of those abstracts and other early efforts, but they started a process that would not have occurred otherwise.

John Bailar became the statistical associate editor for the NEJM in 1981 and supervised much of the detailed work from beginning to end of the project. He recalls how Fred organized the effort:

> The working group met over a period of almost five years at intervals of about two weeks during the school year and one week during the summer. Subgroups developing specific papers and chapters met separately and reported to the main group. These subgroups are largely reflected in the authorship of the project's papers. One subgroup (Bailar plus Thomas Louis, Philip Lavori, and Marcia Polansky), christened by Fred "The Gang of Four," was especially productive and wrote four full-length papers.
>
> Group meetings often went down unexpected byways. I recall discussions of one versus two tails in testing hypotheses, the didactic value of sports statistics (one of Fred's passions), and the latest offering of Shakespeare on Boston television. But when a byway had served its purpose—whether to illuminate a relevant point, or to provide relief from the continuing hard work, or to educate us all on some point of statistics—Fred brought us back to the serious business of getting the writing done. Sometimes, of course, other affairs kept him away. The result was almost invariably a less productive as well as a less interesting meeting.
>
> Fred kept the focus very firmly on the practice of statistics—

> what our study of published papers had to say about current practice, and how this practice of statistics, by imperfect investigators in an imperfect world, might be improved. He insisted that this be carried out by explaining the ideas involved rather than by presenting formulas or computations or by deep study of specific cases. Real examples were used in profusion, but always to illustrate actual practice, bring out the basic ideas, and help readers to understand what was going on so that they could improve their own practice. And positive examples were to be used whenever they could make an important point; Fred sees little profit in the public pointing of fingers at a transgression or a transgressor.

As with most of Fred's other projects, the New England Journal Project resulted in a book, *Medical Uses of Statistics* (B56), which Fred co-edited with John Bailar. It includes citations to some thirty other published products of this working seminar—chapters from the book that were first published in a NEJM series entitled "Statistics in Practice," introductions and chapters in other books, and many papers in other journals. This is an astonishing return on a rather small investment in direct financial support. The catalyst, of course, was Fred Mosteller, who not only conceived and developed the idea, but consistently put more of himself into the project than any other member of the working group. And, as other chapters of the present volume document, at the same time he was just as deeply involved in a broad range of other activities. But, perhaps most importantly, the project changed the statistical quality of papers in the NEJM itself and as a consequence influenced other major medical journals.

A number of features that characterize Fred's approach to "the working seminar" are exemplified by the New England Journal Project. John Bailar describes them in the following way:

> It seems likely that Fred's first thoughts about a possible seminar group have to do with published output. Only when he has that much clearly in mind does he feel equipped to ask for funds, assemble a group, and set them marching toward a common goal while the others are still wondering what it is all about. I do not mean that he has preselected any findings, conclusions, or recommendations—Fred is far too good a scientist for that—but he seems to have an intuitive grasp of what topics are worth study, who should work on them, and how the products, whatever their message turns out to be, should be organized and presented. All is targeted to a clear assessment of some important, real problem.
>
> Nor does Fred ever lose his grasp of essential detail. He goes over each manuscript many times; he checks references to make sure that inferences as well as facts are correct; and he gives

much thought to chapter titles and sequences and to the cross-references among chapters. He surely put more time and effort into *Medical Uses of Statistics* than any other contributor, although his identified contributions to specific chapters are quite limited. Figures and tables must be just so. He checks the consistency of terms, references, and annotations as well as ideas. All the while, he keeps other participants moving toward publication by means of periodic group meetings, notes, telephone calls, and opportune encounters in the hall or elsewhere.

When other members of the working group have finished (they think), Fred becomes even more involved; as others tire and begin to flag, Fred increases the pace of work more and more. His gentle but firm persistence keeps things moving until the manuscript is off to the printer. Then—relax? Never. Fred takes on all the details of production, too, to ensure that the hard work already invested has maximum payoff. He attends closely to such matters as page size and margins; type font and size; the placing of footnotes and other additions to the narrative; the color and design of the cover; and countless other details. He himself reads all the proofs, both galley and page, reviews the corrections to make sure they have been done per instructions, and requires (and reviews) repeat proofs if the work is not satisfactory. His attention to the choice of a book title is legendary. We must have considered a hundred titles for *Medical Uses of Statistics.* Our final first choice, *Statistics in Practice*, appeared on another book after we were well along with final editing, and it had to be replaced. No one who has gone through this with Fred could fail to learn the need for constant attention to detail, or how to exercise that attention.

Where and how did Fred learn to use the working seminar so effectively? I do not know, and this question has clearly puzzled many others as well. "How does he do it?" we ask each other, and find no really satisfactory answers. Certainly there is more to his success than the basic idea, because others who have tried to emulate his approach have generally had substantially less success. The answer must lie somewhere in his unique combination of leadership, method, personal ability, hard work, dedication to detail, and experience. Whatever it is, those of us who have worked with Fred in one of his seminars have learned and grown, and our own future work will be very much improved.

In the Department of Health Policy and Management

Fred Mosteller became chairman of the Department of Health Policy and Management at the School of Public Health on July 1, 1981. There were many good reasons for him to decline another administrative post, but he didn't. Perhaps it was loyalty, or a commitment to a vision of what such a department in such a school ought to be; perhaps it was cussedness, or his sense of humor. Only someone as well-organized as Fred could have even considered the prospect of taking on the leadership of yet another department.

When Fred took charge, HP&M was a department in name only. Its faculty and staff had been divided along programmatic lines—two-year program, executive programs, one-year and doctoral programs. In addition, many members of the Center for the Analysis of Health Practices held appointments in the department. There were 25 faculty members, 14 of them part-time. The department was responsible for thirty percent of all student course credits at the school.

Some changes in the department came fast. Mail got answered. Staff came to have confidence that they had the chairman's attention. As his colleagues always learn, Fred believes in manners and he is a master of gesture and nuance. The budget process was revamped and regularized. Administrative functions were reorganized along departmental lines. These were low-profile steps. They helped the department practice keeping its feet on the ground. Common sense and knowing the score, the department learned, are a foundation for big ideas.

A diverse and sometimes fragmented faculty needed ways to build some intellectual cohesion. Fred arranged to have regular departmental colloquia and, as time went on, the department found additional ways to organize seminars and study groups aimed at building up the department's research. The department took a comprehensive look at its curriculum and, in concert with other units of the School, paid particular attention to international health. Moreover, Fred stressed the importance of a strong doctoral program, through which the strengths and intellectual work of the department would be perpetuated. The fields of health decision sciences and health economics were significantly strengthened with new faculty appointments and new course offerings. Stephen Thomas recalls:

> During all these efforts, nothing Fred did exceeded the impact of his simply being himself. New departmental priorities were not so much declared as personified. Fred's new colleagues found their old habits and expectations laid bare by his own level of commitment and his own prodigious productivity. There seemed to be no explanation for how he could accomplish what he regularly accomplished. He was a constant reminder that there are no tricks, only gifts and the burden of hard work that gifts impose.

Along the way, Fred also took on the chairmanship of the School's Committee on Educational Policy, the group responsible for oversight and review of the School's teaching programs. Under his leadership, the faculty adopted a new program to train physicians in clinical effectiveness, a new extension school program in public health, and a systematic approach to reviewing departmental programs.

It took Fred several years to get a display case hung to show off recent faculty publications—and he did not choose to keep it full singlehandedly. Fred referred to the display case as his most tangible accomplishment as chairman. (In moments of great frustration, he has been heard to call the case his *only* accomplishment.) But there were many others. As chairman of the department, Fred was always the teacher to his colleagues. "He sets the example," one staff member has said. "He opens the door and he expects no less of himself than he expects of others. He inspires responsibility in them, and their shortcomings he sometimes feels as his own."

7.6 The "Statistician's Guide to Exploratory Data Analysis"

Simultaneous with all these activities in the School of Public Health, Fred continued to undertake major projects that were based in the Department of Statistics, supported primarily by a sequence of grants from the National Science Foundation. One project, known internally as KAP (Knowledge Acquisition for Policy), produced the book *Data for Decisions* (B44), already mentioned in Section 7.4. The largest of the projects had the acronym SGEDA for Statistician's Guide to Exploratory Data Analysis.

The SGEDA project set out to approach exploratory data analysis (EDA) from a perspective different from that in John Tukey's *Exploratory Data Analysis* (1977) and the Mosteller-Tukey "green book," *Data Analysis and Regression* (B42). As the overall title "Statistician's Guide" suggests, the aim was to discuss the conceptual and theoretical background for a variety of techniques of EDA and to show how they are related to more traditional techniques in statistics. The project got under way in 1976, somewhat earlier than the NEJM project, but following a similar organizational model. The three editors (Fred, John Tukey, and David Hoaglin) held extensive planning meetings, mapped out the overall objectives, and developed an outline for three volumes. From among graduate students, faculty, academic visitors, and others interested in EDA, they recruited a team of participants and began to examine possible assignments and the chunks of research that would be needed to provide a thorough account. The group met weekly to discuss progress, problems, and plans and to share and criticize a steady stream of memos and drafts. Aside from the immediate results of the work, perhaps the most important thing the participants learned was how to give and receive criticism on manuscripts. A cohesive group with a common goal made this easy, and on several occasions Fred specifically discussed the art of giving constructive criticism: First, tell the author how wonderful the manuscript is; then explain your ideas for making it even better. From this effort came the two volumes of the "Statistician's Guide": *Understanding Robust and Exploratory Data Analysis* (UREDA for short), published early in 1983 (B47), and *Exploring Data Tables, Trends, and Shapes* (EDTTS), published in 1985 (B54). Although each chapter carries the name(s) of its author(s), all parts of these books had the benefit of heavy central editing, so that they appear almost to have been written by a single person. All three editors contributed extensively to this process, and they also read all the galley and page proofs. Throughout the process Fred's gentle but pervasive leadership fostered pride in workmanship.

Even before the completion of UREDA and EDTTS, the scope of the weekly meeting broadened to include other projects and research activities support by the NSF grant or by contracts from the U.S. Army Research Office. Most participants appreciated the benefits of having a friendly group of colleagues available as a sounding board for ideas and as readers of drafts. Later the group served as a ready set of trial subjects for a questionnaire that would investigate the numerical meanings that people associate with various probabilistic expressions (a follow-up to the results for medical professionals reported in P156, it led to P180 and P182). In the meetings, Fred often reported on the progress of his other projects, such as the NEJM project. Even so, from his high level of activity within the group, few participants would guess that the other projects, at the School of Public Health and elsewhere, could possibly be so extensive.

These team efforts seem to be a characteristic part of Fred's style, if not unique to it, and they have enabled him to deliver a rich payoff for many scientific enterprises. For their sustained help in riding herd on many of the projects, he would be quick to thank his staff in the Department of Statistics: Cleo Youtz and Marjorie Olson, Fred's secretary since 1972.

7.7 Epilogue

Harvard clearly has had a substantial impact on Fred Mosteller, but Fred's impact on Harvard has been profound, extending into virtually every part of the University. Fred has chaired four departments and untold University committees. He has organized courses, working seminars, and major scientific projects. But throughout his over 40 years on the faculty he has always made time for students and colleagues—time to answer a statistical question, or time to collaborate on a paper or project. So many of us have worked on projects with Fred at Harvard and become his coauthor or collaborator that our names would fill a separate book.

The spirit of Fred's influence on his colleagues was captured by a School of Public Health colleague in the following poem, read at a dinner in Fred's honor in December 1986.

THE LAST COAUTHOR

by

Donald M. Berwick

On a high and secret mountain on a South Pacific isle
Lived a hermit in a mud house in a most reclusive style.
He had not clothes nor money, neither dishes nor a bed.
And he had never even written one short monograph with Fred.

Such a thing had not been heard of; it had not before appeared.
And so, Fred climbed the mountain since he thought it was so weird.
"Welcome, pilgrim," said the hermit. "Let's get to work," said Fred.
"I think not," said the hermit. "Let us meditate instead."

"That's fine," said the Professor without the slightest frown,
"But, while meditating, why don't we just write it down?
Soon we'll have two hundred pages, and a publisher we'll seek
For a book by hermit, Mosteller, and maybe Bunker and McPeek."

The hermit felt suspicious and he gave a cautious look.
"Why would you want," he asked, "to write a Meditation book?"
"I'm itchy," was the answer. "Things are getting very slow.
The last book that I finished was a whole three days ago."

"But, I don't know statistics." Answered Fred, "Oh, is *that* it?
In the face of gaps in knowledge *other* fakirs haven't quit.
From your acquaintance with Nirvana other expertise can stem.
I wrote textbooks on statistics; they gave me H, P and M.[1]

"Come, now, let's write together; to a book we must give birth.
I've written fifteen papers with everybody else on earth.
You're the last unclaimed co-author. Hermit, friend, don't be a meany.
Collaborate with me and you'll have perpetual zucchini.

I'll send postcards to you daily with perspicacious quips
From the countries that I visit, if I ever do take trips.
I will teach you to hold meetings in which everyone takes part.
And decides to do exactly what you wanted from the start."

The hermit got a modem, and the rest is history.
The book they wrote each child hears from every father's knee.
By hermit, Ware, and Mosteller (edition number ten):
The Health Care Risks of Jackknifing with Cost-effective Zen.

---

[1] Health Policy and Management

The hermit now has tenure, though he kept his old mud place
When Harvey[2] said that Kresge was completely out of space.
And, he and Fred both meditate whene'er they're both in town,
Though Fred insists their mantras must be always written down.

The hermit thought it silly that his friend would take this stance
'Til Fred explained that goal he had: to learn the laws of chants.
The hermit, Ware, and Mosteller then settled in anew
To write the book that soon became their opus number two.

And three and four did soon emerge from his impatient pen.
When Fred has finished any book, his next word is, "Again."
But now, Fred, for one evening, we will resist your call,
And instead say, "Happy Birthday," from your worshipful "*et al.*"

Appendix: Excerpts from "Some Topics of Interest to Graduate Students in Statistics" by Frederick Mosteller (dated April 21, 1970)

The following remarks are intended to be of some help to beginning and continuing graduate students. Times change, and facts can go out of date rather quickly.

Colloquium

This activity is part of the graduate student's education and of the continuing education of the faculty.

First, graduate students are expected to help with colloquium. Every professional has duties in connection with his organizations.

Second, you should attend the colloquium regularly. Sometimes you will have no idea what the speaker is talking about. Find out why. Is he giving a bad speech or is the material just too advanced? One way to avoid having no idea what the speaker is talking about is to find out the topic in advance and read up a little before going, or ask someone about it.

Third, meet the speaker. Go to the tea and meet him beforehand if you can. Then after chatting with him a little bit let someone else meet him too.

Fourth, ask a question if you can. Part of the courtesy shown a speaker is in the audience's interest. If you are chairman you are especially respon-

[2]Harvey Fineberg is the Dean of the Harvard School of Public Health.

sible. The chairman should prepare a couple of people beforehand to ask questions so that if no one rushes forward voluntarily, the discussion can begin with their questions. Furthermore, as chairman you should have at least one question in case all other persons fail. A paper without discussion reflects on the audience and makes the speaker very sad.

A chairman has to know how to stop the discussion gracefully too. After a good discussion there may be a brief pause, and this gives the chairman a chance to thank the speaker and end the meeting. Don't make it drag out by hunting for questions that aren't there.

Fifth, if he is not already surrounded, go up afterward and thank the speaker for the talk. Pick something specific to mention. You may learn something interesting.

Sixth, plan to speak at colloquium.

Tasks Outside Class

You will occasionally be asked to do odd jobs of a professional kind with no thought of there being any remuneration. You may be asked to help a student from another department as a consultant. You may be asked to work over a paper someone has submitted to a journal as part of the refereeing process. You might be asked to solve a small problem that someone needs the answer to but hasn't time to work out. These little chores are part of one's education. If no one asks you to do something, perhaps you should take some professor as a target and go ask him if there isn't something you can do.

Teaching Fellowships

The department anticipates that most people will at one time or another hold a teaching fellowship and a research assistantship. The teaching fellow's duties vary depending on the course he works in. In one course he might have some or all of the following duties.

0. Attend the lectures.

1. Attend meetings planning the material, including term paper projects, preparing drafts of examinations, midterm and final, proofing them, assisting with the lecture in some way—perhaps a demonstration.

2. Preparing programs and problems for computer work.

3. Holding a session a week with a class of students.

4. Giving an occasional lecture.

5. Holding office hours.

6. Grading papers.

7. Preparing bibliographic materials to help students get started on a project.

8. Grading midterm and final examinations as well as term papers, and preparing final grades.

It is important that the TF be loyal to his course. When there are several ways to do a thing and the choice of way has been settled at the section leaders' meeting, then do it with a will. When you find some topic going badly, it is your job when the section leaders meet to bring it up with some concrete proposals for remedy. Loyalty doesn't mean blindness. It means cooperation.

When you find yourself getting in over your head on teaching, talk it over with the person in charge of the course. Teaching can grow to be all consuming. If you take on too many jobs, then you can find yourself neglecting your other duties. But when you take on a job, do it and do it on time. A course can't wait; it moves inexorably.

Now some points to remember.

Start your class on time. Seven minutes after the hour is as near as there is to a clear rule about this. Holding your class is regarded as important at Harvard, and missing one is not regarded as a joke. If you are sick, then some plan needs to be made to take care of the class.

Have enough material to use the whole hour. Students may seem to like it when they are let out early, but what they use the time for is to report to the instructor in the course that the TF doesn't know what he is doing.

Sessions with the instructor in charge of the course should be attended on time. In case of illness send word.

Hold your office hours sacred, and be where they are supposed to be. Don't create confusion by changing the time and place once it is set. It is impossible to get the word around. Stay until the end even if no one comes.

Even if you are very talented, you will be wise to make sure you can do all the problems before a problem session.

Grade your homework papers carefully. What is important here is not so much the grade itself but what you write to the student on his paper. Better not make smart remarks. Tell him how to do problems he has trouble with right on the paper. Get them back soon.

If during the class you think you haven't gotten something across, begin asking the students some questions. It will help root out the trouble. Just don't expect too much of them.

You cannot excuse anyone from an examination.

A teaching fellow who does a bad job gets bad notices in a book put out by undergraduates. Unfortunately, so do some who do a good job.

Research Grants and Contracts

Research in the academic world is largely supported by grants and contracts made with the government or with private foundations. ... [The principal investigator] has the responsibility to do the research and to report regularly on progress. He does this partly with regular reports and partly by submitting reprints of articles the project has published. The agency will tell him what kind of reporting it wants, and how often.

Thus the product is research papers and the ideas in them.

The student who is working on a research assistantship is therefore supposed to be contributing quite directly to the product of the project, and finished research is this product.

What if nothing works? That would be a rather black mark against the investigator. You can readily see that not much research support will be forthcoming for investigators who don't produce.

Research Assistantships

These positions are paid jobs. The student will ordinarily wish to spend more than the quoted amount of time on them because such research work contributes to advancement toward the degree and to long-run professional accomplishments.

The research assistant's prime responsibility is to do research. By and large it is expected that it will be in the direction that a project is working. Naturally we all get ideas for research that is not directly on the topic a project is devoted to. A certain amount of this "outside the program" research is perfectly acceptable in the report of work done under the program.

The research assistant is expected to keep in close and regular touch with the project director. And, most important, he should write things up as he goes along. So far as the program is concerned, research that is not written up is research not done.

When the research assistant plans to take a vacation, he should not just vanish but work it out with his project director. For most academic research workers vacation time means research time.

Manuscripts

The objective in writing papers is that people will read them. In planning a paper, ask yourself "For whom am I writing?" Try to catch and hold the reader's interest. He is busy and easily distracted into reading something else more interesting. If you feel that a complex or dull digression is needed from your main theme, tell the reader when you start this why it is essential. A brief, informative summary at the beginning helps greatly, and more and more journals are insisting on this.

In drafting a manuscript, put it away for a week or two after you first write it. Then you read it more nearly from a reader's point of view in preparing the second draft. Writers often fail to define symbols and technical terms that they use.

Plan on writing several drafts. Only a remarkably gifted author can write a "final version" on the first or even second try.

Theses

Read the advice for preparation of theses. Ask the Department secretary for a copy. Don't just ask around.

Be sure to get several copies of the thesis made. You will need them. You will want to present a copy to your advisor and to others who have helped you. Some friends will want one. You may wish to give copies to prospective employers. A project supporting your work will need copies. You may need a few copies to cut and paste in writing up articles. Don't try to save money on copies of the thesis.

Buying Reprints

If you publish an article, buy reprints. Although it costs money, it is part of your business. The few reprints provided by the journal will not last through your career.

Giving Talks

After doing research, you will want to present it. The two main ways are publication in professional journals and talks in seminars, colloquiums, and at regional and national meetings of professional societies.

You will be expected to talk here at the Department as part of your preparation. When you are being considered for a position elsewhere, you will ordinarily be asked to speak on your latest research. Whether you are looking for a position or not, you should regard each of these talks as bearing upon your professional future.

The first rule is, talk only when you have something to say, and the second is, when you talk, say it well.

Preparation

Although a few great speakers may be born, most are made. This is especially true of technical presentations. At a contributed papers session of a professional meeting, one may be allotted 10 minutes to discuss a very difficult topic. Only great self-restraint and considerable preparation can make such short presentations successful.

In the mathematical sciences we tend to think of blackboard and chalk as the primary visual aids. But people who rely heavily on them in public lectures rarely do as well as they could.

The Handout

People like to take something away from a talk. If you have a handout, you have already made sure that they will. You have also taken out insurance against a major omission in your presentation. You have prepared for amplification in case of questions. You have put the listener in the position of being able to go at a slightly different pace than you, and he can go back and check the definitions and conditions in a leisurely manner. You have also shown that you cared enough about the audience to take the trouble to prepare.

Try to find out about how many handouts you need and take plenty. Do not depend on mail. Be sure your handout has a title, a date, your name and affiliation, reference to support if any (NSF, NIH, Rockefeller, Ford, etc.), numbered pages. Some speakers include a reference to the place where the talk is being given. Beyond this, a lot depends upon the topic, but key definitions, theorems, proofs, formulas, references, tables, figures, and results are examples of valuable things for the reader to carry away and for you to lean on in your presentation. You need not hand out the whole paper.

While there are occasions when one wants to build a complicated formula step-by-step at the board, the act of writing a 3- or 4-line formula is wearing for the arms and the viewers, and frequently leads to mistakes. Board space is often limited. Turning your back on the audience to write decreases audience interest.

Troubles

In giving speeches away from home you have to be prepared for the possibility of poor facilities. A person with a handout is ready when they say "We had no idea this place had no board; they hold conferences here all the time."

Before a talk go to a bit of extra "trouble." Find out where it is to be held. Go there and try everything out. See if you can turn the lights on and off. If there is a projector, be sure you can plug it in and that it works. Focus it. In the case of an overhead projector encourage them to provide an extra bulb in case the bulb fails.

If there is audio equipment, try it out. Find out how far you should be from the mike to make the right sort of noise. Have some water nearby in case your throat goes dry. A sip clears up lots of voice troubles for a speaker.

You may feel embarrassed about all these preparations, and a person testing a mike always feels silly. But you'll look even sillier fumbling around during the talk.

If you haven't enough handouts, get the ushers to distribute one to every other person.

Don't distribute the handouts yourself. The speaker has other things to do.

Strange things happen. Be ready for them.

1 The room where the talk is to be given is occupied. Relax. Let the local people work it out. Introduce yourself to people waiting around and chat with them. Don't complain.

2 The room isn't big enough! Already you are a great success. Don't apologize for the hosts. You can say how happy you are to see such a large audience. Don't apologize for late starting because of local arrangements. Only apologize for lateness if it is your fault.

3 Only two or three people show up. Frequently there will then be a dither about whether the speech should be given or not since there are so few. Indicate that you would like to give the speech and give it, perhaps a little more intimately than otherwise. Don't talk about the smallness of the audience.

Projectors

There are many kinds of projectors, but three stand out—movie, overhead, and slide. ... Overhead projectors are easy to use and the overlays can be easily prepared even by an amateur. Time and trouble are repaid for goodlooking overlays, though. This projector is a great supplement for a handout.

One can write with a [special marker] on the overhead projector's overlay instead of on a blackboard, but it takes practice.

The speaker may not realize how much small movements are magnified by the projector. When pointing to something with a [marker], rest the tip on the plastic to stop the quivering.

There are many tricks with projectors. The main rule is, write large and don't put much on a page.

Rehearse

The idea of rehearsing one's speech may seem strange to those who have never seen speakers still defining their opening terms when the chairman declares the session concluded. Mathematical talks frequently have a natural flow, but the researcher usually knows so much more than he has time to tell that he has to exercise great restraint. This restraint can be aided by rehearsing the talk in full with all props treated exactly as they will be in the speech.

In rehearsing, if you plan to write on the board, go ahead and write, but include everything in the time of the speech.

Often it turns out on rehearsal that the speech is at least twice as long as the speaker planned. Furthermore, the actual delivery will likely be slower than the rehearsal. One's first impulse is to try to speak faster. This is fatal, forget it. Second impulse, go through the talk, if it has been written out, and wipe out sentences here and there. This usually helps little.

The basic rule is that a speech that is too long has too much in it. You must kill whole chunks. Rank order the things of importance. Is it the method of proof or the result that is important? The formula or the table? Decide this and cut to suit. Remember that we rarely hear speeches that are too short for the audience. Your handout can help you with this cutting.

Use your watch and have a manuscript marked with times so that you can tell how nearly your speech is following its schedule. Be prepared to dump something.

Good speakers differ a lot in their devices. Some merely follow an outline. Some actually memorize their speeches. Some use manuscripts to keep themselves in line. Some read their speeches. (A committee report may word things very delicately.) Reading is usually not a good idea, but some occasions require it. The main thing is to rehearse with a clock a few times until one can "hear what one is saying" instead of just struggling with the content.

Papers at Professional Meetings

Two main types of papers occur at meetings: contributed and invited. Some societies are very strong on having contributed papers sessions. They announce them well in advance, and there may be some small thing you have to do to get on the session, for example submit your name and an abstract of your paper by a given time. Generally speaking, these sessions are easy to get into. The presentation of the paper itself is allowed very little time, different times at different meetings. The shortest the writer has seen was 6 minutes per paper, the longest 20. The time is somewhat determined by the number of papers submitted. If you want to contribute a paper at a meeting, then look through the journal for the meeting announcement and calls for papers.

To be invited to give a paper, someone has to know that you are working on a topic, and there likely has to be a session being arranged on that topic. It is somewhat chancy. The pipelines into the program committee and to the chairman who is arranging a particular session are usually fairly personal. Sometimes a professor in a department will be on an organizing committee and can help. ... One way to get in on such sessions when you have something to offer is to become acquainted with the other people working in your field by attending meetings where they will be speaking and introducing yourself, by getting your advisor to put you in touch with people, and by picking up your pen and writing a letter to the people you want to know and whose work interests you. A letter asking for a reprint is a good opportunity to tell of your own interest and work in brief. Similarly, you can distribute your own reprints to people interested in your field.

References

Anderson, S., Auquier, A., Hauck, W.W., Oakes, D., Vandaele, W., and Weisberg, H.I. (1980). *Statistical Methods for Comparative Studies: Techniques for Bias Reduction.* New York: Wiley.

Birch, M.W. (1963). Maximum likelihood in three-way contingency tables. *Journal of the Royal Statistical Society, Series B*, **25**, 220–233.

Bunker, J.P. (1972). *The Anesthesiologist and the Surgeon: Partners in the Operating Room.* Boston: Little, Brown.

Bunker, J.P. and Blumenfeld, C.M. (1963). Liver necrosis after halothane anesthesia. *New England Journal of Medicine*, **268**, 531.

Deming, W.E. and Stephan, F.F. (1940). On a least squares adjustment of a sampled frequency table when the expected marginal totals are known. *Annals of Mathematical Statistics*, **11**, 427–444.

Haberman, S.J. (1974). *The Analysis of Frequency Data.* Chicago: University of Chicago Press.

Neutra, R., Fienberg, S.E., Greenland, S., and Friedman, E.A. (1978). Effects of fetal monitoring on neonatal death rates. *New England Journal of Medicine*, **299**, 324–327.

Parsons, T. (1956). *Department and Laboratory of Social Relations, Harvard University, The First Decade, 1946–1956.* Report of the Chairman. Harvard University.

Tukey, J.W. (1977). *Exploratory Data Analysis.* Reading, MA: Addison-Wesley.

Virginia and Fred Mosteller at Virginia's family summer home in Butler, Pennsylvania, about 1937.

Virginia and Fred, wedding picture, Princeton, May 1941.

Fred with a counting sorter, working for Hadley Cantril, 1940.

Fred working at Harvard, about 1952.

Fred working with an electronic calculator, 1955.

Fred working with a slide rule, 1955.

Original faculty of the Harvard Department of Statistics, May 1959. From left to right: John Pratt, Howard Raiffa, William Cochran, Arthur Dempster, and Fred.

Fred with mechanical urn sampler, 1964.

Fred learning to ice skate in Belmont, Massachusetts, 1964.

Fred at his summer home on Cape Cod, 1976.

Fred receiving honorary degree from President Richard Cyert at his alma mater, Carnegie Mellon University, 1974.

Fred with colleagues in the Department of Statistics, University of Chicago, after receiving an honorary degree, 1973. From left to right: Leo Goodman, Paul Meier, Fred, David Wallace, and Patrick Billingsley.

Fred with Colin Campbell, President of Wesleyan University, on his left and Stanley Lebergott on his right, on the occasion of receiving an honorary degree, 1983.

Chapter 8

REVIEWS OF BOOK CONTRIBUTIONS

Edited by
David C. Hoaglin

Introduction

A look at Fred Mosteller's bibliography reveals that he has had a hand (usually two) in well over fifty books—as author, coauthor, coeditor, or collaborator. Many of the titles are immediately familiar, either in statistics or in a number of other fields (as we should expect from the portrait of Fred as a scientific generalist—Chapter 3).

Besides their diverse subject matter, the books vary in form and level. Although no simple scheme quite suffices, many of the books fit into three broad groups:

Texts and related expository works

Probability with Statistical Applications (1961, 1970)
Statistics: A Guide to the Unknown (1972, 1976, 1977, 1978, 1989)
Sturdy Statistics (1973)
Statistics by Example (1973)
Statistics and Public Policy (1977)
Data Analysis and Regression (1977)
Beginning Statistics with Data Analysis (1983)
Biostatistics in Clinical Medicine (1983, 1987)

Monographs

Sampling Inspection (1948)
Stochastic Models for Learning (1955)
Applied Bayesian and Classical Inference: The Case of the Federalist Papers (1964, 1984)
Fifty Challenging Problems in Probability with Solutions (1965)
Data for Decisions: Information Strategies for Policymakers (1982)
Understanding Robust and Exploratory Data Analysis (1983)
Exploring Data Tables, Trends, and Shapes (1985)
Medical Uses of Statistics (1986)

Reports of panels, committees, and working groups

The Pre-election Polls of 1948 (1949)
Statistical Problems of the Kinsey Report (1954)
The National Halothane Study (1969)
Federal Statistics (1971)
On Equality of Educational Opportunity (1972)
Weather & Climate Modification: Problems and Progress (1973)
Costs, Risks, and Benefits of Surgery (1977)
Assigned Share for Radiation as a Cause of Cancer (1984)
Assessing Medical Technologies (1985)

Even though the lists under these headings aim to be illustrative rather than exhaustive, one cannot avoid noticing their diversity. These books and others have educated, informed, and stimulated a wide range of audiences.

Interestingly, each list starts early in Fred's career and continues today. Indeed, some of the most recent books appeared after we had begun to assemble the reviews in this section, and others are making their way toward publication as this introduction is being written. We look forward to the appearance of those and others in the years to come.

To highlight the breadth and depth of impact of the books in which Fred has participated, we commissioned new reviews of about twenty of the books. We encouraged the reviewers, besides approaching these contributions in the usual scholarly way, to discuss a book's influence, both on themselves and on others in the field. They responded generously with a variety of engaging reviews. Many reviewers were able to draw on their experience as Fred's colleagues or students and have thus enriched their accounts with personal anecdotes that offer further glimpses into his styles of teaching, research, and writing. In all these ways, the reviews that follow add a further dimension to the enduring impact of Fred Mosteller's contributions to statistics, science, and public policy.

Sampling Inspection

H.A. Freeman, Milton Friedman, Frederick Mosteller, and W. Allen Wallis, editors, and Statistical Research Group of Columbia University. *Sampling Inspection.* New York and London: McGraw-Hill, 1948.

by
W. Allen Wallis

Fred Mosteller is an accomplished magician and prestidigitator, and an acute analyst of the psychological interactions between performers and observers. While I had enjoyed those talents on informal occasions, I did not appreciate their pedagogical value until with Fred I participated in the team that prepared *Sampling Inspection: Principles, Procedures, and Tables for Single, Double, and Sequential Sampling in Acceptance Inspection and Quality Control Based on Percent Defective*, usually cited simply as *Sampling Inspection* or even just *SI.*

SI carries a publication and copyright date of 1948, but an earlier version was written and distributed in "Ditto" form—"Ditto" was a brand of duplicator then widely used—in the three spring months of 1945. Then, early in June, it was used as the text for a five-day course at Hershey, Pennsylvania attended by "inspection authorities from all naval inspection districts ... and from technical bureaus concerned with procurement." Fred was one of the principal teachers in that course, and it was there that I saw his skills as a magician produce a superb pedagogical performance. Not only did he focus the attention of his students precisely where he wanted it, but, just as important, he kept their attention away from complications until he was ready to deal with them. He was able in this way to simplify the essence of a complicated matter and get it understood, then—before it was permanently imprinted in his hearers' minds in oversimplified form—to move their attention to one qualification, modification, or elaboration after another until they understood thoroughly some fairly complex matters.

So many members of the Statistical Research Group[1] contributed to *SI* that its authorship was ascribed to the Group as a whole. The bulk of the actual writing of the 1945 version, however, was done by Milton Friedman, Fred Mosteller, and Jimmie Savage, with Milton in charge. (Harold Freeman was responsible later for converting the 1945 Navy manual into the 1948 McGraw-Hill book.) They worked under extraordinarily high pressure, since most of the three months was used preparing tables and charts, working out procedures, and—finally—typing and reproducing the manual. At that time neither Fred nor Jimmie had attained standards of style

---

[1]For an account of the Statistical Research Group, see W. Allen Wallis, "The Statistical Research Group, 1942–1945," *Journal of the American Statistical Association,* 75 (1980), 320–335 (with discussion).

and precision comparable with Milton's, and they usually came away from the writing sessions with Milton in a state of extreme frustration and exasperation. As I said in a memoir about Jimmie,[2] probably they hoped they would never again lay eyes on Milton. In large part because of this experience, Fred and Jimmie both became extraordinarily fine writers, not surpassed by anyone in their fields. (All three became lifelong friends and mutual admirers.)

Fred is mentioned in the Preface to *SI* as one of six people who prepared the manuscript of the Navy manual out of which *SI* came, and as one of four editors who "planned, directed, and reviewed" the work. To him and David H. Schwartz is ascribed primary credit for the two chapters on process control.

How does *SI* look in the light of 40 more years in the history of statistical sampling inspection? None of the military standards that evolved from *SI* (which itself evolved from earlier military standards, which in turn had evolved from Bell Laboratories standards) have benefited from theoretical sophistication in economics and statistics equal to that of Friedman, Mosteller, and Savage. On the other hand, both economic theory and statistical theory had advanced by the time later standards were prepared, and some of the later standards were influenced by excellent statisticians, especially Albert Bowker (himself a member of the Statistical Research Group) and his colleagues at Stanford. Practice, also, has progressed; for example, in many cases measurements of quality have been automated to such an extent that their cost is negligible and one-hundred-percent inspection has replaced sampling.

For a historian of statistical theory and practice or a historian of industry and technology, statistical sampling inspection offers an attractive field of investigation, as yet little cultivated. When at last a thorough history is written, the work of Fred Mosteller will occupy a significant place in it.

---

[2]at pp. 11–24 in *The Writings of Leonard Jimmie Savage—A Memorial Selection.* Washington, D.C.: The American Statistical Association, 1981.

The Pre-Election Polls of 1948

Frederick Mosteller, Herbert Hyman, Philip J. McCarthy, Eli S. Marks, and David B. Truman, with the collaboration of Leonard W. Doob, Duncan MacRae, Jr., Frederick F. Stephan, Samuel A. Stouffer, and S.S. Wilks. *The Pre-election Polls of 1948.* Social Science Research Council, Bulletin 60, 1949.

by
Richard F. Link

The American public awoke on Wednesday, November 3, 1948 to learn that Truman had beaten Dewey and would continue as President for four more years.

Shock and consternation greeted this news, because the polling organizations had led people to expect that Dewey would prevail. Dismay was especially strong among social scientists who used polling techniques. Mindful of the *Literary Digest* fiasco only twelve years earlier, they feared that their more scientific polling methods would also fall into general public disrepute.

Therefore, the Social Science Research Council quickly formed a committee, headed by S.S. Wilks and Frederick F. Stephan, to investigate what had gone wrong. The committee appointed a technical staff to conduct the investigation, with Frederick Mosteller as chief of staff and members Herbert Hyman, Philip J. McCarthy, Eli S. Marks, and David B. Truman.

The quick action of the SSRC, the cooperation of the major polling organizations, and the almost incredible speed of the technical staff resulted in a published report on December 27, 1948, only five weeks after the formation of the technical staff. Although controversy over the polls continued, the committee's report and the subsequent SSRC Bulletin 60, entitled *The Pre-Election Polls of 1948*, which was published in 1949, provided an authoritative basis for assessing the strengths as well as the weaknesses of then-current polling practices.

We focus here on two aspects covered in Bulletin 60: the general conclusions about polls contained in the December 27, 1948 technical staff report, and the discussion of measuring the error, written for the Bulletin by Fred Mosteller.

The December 27, 1948 report gave a seven-point summary of conclusions (pp. 290–291):

1. The pollsters overreached the capabilities of the public opinion poll as a predicting device in attempting to pick, without qualification, the winner of the 1948 presidential election. They had been led by false assumptions into believing their methods were much more accurate than in fact they are. The election

was close. Dewey could have won by carrying Ohio, California, and Illinois which he lost by less than 1 percent of the vote. In such a close election no polls, no advance information of any kind, could have predicted a Truman or Dewey victory with confidence. The failure of the polls was due to neglecting the possibility of a close election and the necessity of measuring preferences very accurately just before the election to determine whether a flat forecast could be made with confidence.

2. The pollsters could have foreseen the possibility of a close contest had they looked more carefully at their data and past errors. They acted in good faith but showed poor judgment in failing to apply in 1948 what they knew about their past errors, and failing to ascertain late campaign shifts.

3. The over-all operation of making election predictions from pre-election polls is a complex one involving eight major steps at each of which error may enter. It is very difficult to unscramble the total error and allocate components of it to these various steps. The evidence indicates that there were two major causes of errors: (a) errors of sampling and interviewing, and (b) errors of forecasting, involving failure to assess the future behavior of undecided voters and to detect shifts of voting intention near the end of the campaign.

4. These sources of error were not new. While Gallup and Crossley were more successful in picking the winner and his electoral vote in 1940 and 1944 than in 1948, their average errors state by state were at least half as great in the two preceding elections as in 1948. Hence, it is possible that their errors in 1948 were due to much the same causes as those that produced the earlier forecast errors, but if so, these causes operated more strongly. Roper's wide discrepancy in 1948 cannot be explained so readily in terms of factors present in his earlier close estimates of the national vote. It appears to be due to the upsetting of the balance previously maintained among various factors in his polling operations.

5. To improve the accuracy of pre-election poll predictions satisfactorily it is necessary to reduce the error at every step in the over-all polling process. The error at some of the steps, notably sampling and interviewing, could be reduced by using methods now available. But reduction at other steps depends on further basic research in psychological and political behavior.

6. The manner in which the the pre-election polls were analyzed, presented and published for public consumption contributed materially to the widespread misinterpretation of the results of the polls and to the great public reaction to their failure to pick the winner. This led to a poor understanding of the lack of accuracy of the polls and of the nature of the errors residing in the polls, with the result that the public placed too much confidence in polls before the 1948 election and too much distrust in them afterwards.

7. The public should draw no inferences from pre-election forecasts that would disparage the accuracy or usefulness of properly conducted sampling surveys in fields in which the response does not involve expression of opinion or intention to act. There are more appropriate methods to check the accuracy of such surveys.

Briefly, these conclusions state that the 1948 pollsters were guilty of "hubris," that they said too much on the basis of too little information, that they ignored their past data problems and overlooked clues in their present data which portended real trouble. This last problem, ignoring one's data, was, of course, present in abundance in the *Literary Digest* material of 1936.

The committee's recommendations for further study remain relevant, even after nearly forty years. Their seven recommendations are (pp. 292–294):

1. To improve the accuracy of polls, increased use should be made of the better techniques now available, particularly in sampling and interviewing. Since the reduction of any part of the error greatly increases the chances of a successful forecast, the committee urges that pollers exert every effort to adopt more reliable techniques.

2. Increased attention should be paid to development of research on each step of the polling operation to attempt to improve methods used in opinion research. This would include research on sampling methods, interviewer bias, concealment of opinions, selection and training of interviewers, etc. Experimental studies of such problems should be planned well in advance of elections. Many of these experiments can be incorporated into regular survey operations and as a part of more general cooperative studies of specific communities.

3. Research should be expanded on the basic sciences, particularly social psychology and political science, which underlie the analysis of voters' behavior. Even if perfect sampling of individuals

is employed, we now know too little about voting intentions, factors affecting change in opinion, prestige effects, and similar topics to predict who will translate his opinion into actual voting.

4. In view of the increasing amount of emphasis being placed on public opinion polls, the committee considers it very important for the public to be effectively informed about the limitations of poll results so it can interpret them intelligently. It urges that polling organizations, newspapers and magazines, and social scientists who work with poll results help provide the public with more information about polls, their interpretation and limitations.

5. There should be more effective cooperation between research workers interested in opinion measurement on common problems of methodology, underlying theory, and research design, including more studies of validity and reliability than have been made heretofore. Such cooperation would benefit all parties through the wider dissemination of their findings, and more efficient use of research resources and opportunities for experimentation.

6. More extensive training facilities and opportunities for practical experience under effective supervision should be provided for students who may be interested in research careers in this field, which includes political behavior, psychological research, and opinion measurement and statistical methods.

7. Analysis of the elections is greatly hampered by long delays in the reporting of official returns. The present inadequacies with respect to the collection of election statistics should be remedied by effective organization for rapid and accurate reporting of election statistics at the local, state, and national levels.

Taking the recommendations in order, we may comment:

1. Certainly most polling organizations utilize sound sampling methods today, but the quality of their interviewing practices appears to vary considerably from organization to organization.

2. Many of the problems alluded to here are still alive and well.

3. Voter screens and their consequences are much better understood today. However, the other points are still viable research issues; see, for example, *Surveying Subjective Phenomena*, Vols.

1 and 2, edited by C.F. Turner and E. Martin, Russell Sage Foundation, New York, 1984.

4. The problem of educating the public, both lay and editorial, about survey error and its importance is still a mostly unwon battle.

5. The American Association for Public Opinion Research and other organizations continue trying to promote cooperation in the dissemination of information, but much remains to be done.

6. Opportunities in academia abound today, but they vary in quality and often lack a "real world" focus.

7. This point has largely been solved by the News Election Service, an organization supported primarily by the broadcasting networks and wire services. The NES reports all national election results in some detail and has been functioning since the fall of 1964.

Chapter V of Bulletin 60, entitled "Measuring the Error," first indicates that "errors" can be measured in many possible ways, enumerating (p. 55):

1. Measure the error in an estimate by the difference in percentage points between the predicted Democratic (Republican) proportion of the total votes cast and the actual Democratic (Republican) proportion.

2. Measure the error in an estimate by the difference in percentage points between the predicted Democratic (Republican) proportion of the two-party vote and the actual proportion Democratic (Republican) of the two-party vote.

3. Measure the error by averaging the deviations in percentage points between predicted and observed results for each party (without regard to sign).

4. Use the concept of average percentage error, taking the ratio of predicted to actual proportion, and averaging the deviations from 100 percent.

5. Use the difference of the oriented differences between predicted and actual results for the two major candidates.

6. Use the maximum observed difference in percentage points for any party.

7. Use the chi-square test.

8. Use the electoral vote predicted versus that observed.

After discussing how these various measures can be employed, including the ease of understanding the various measures, Fred considers specifics relating to pre-election poll results. In particular, he highlights the presence of a rather substantial and persistent Republican bias in many poll results in past elections. One important implication of this phenomenon is the effect that such biases, even relatively small ones, can have on electoral college estimates. This serves to remind the reader that presidential elections are decided in the electoral college and that a bias as small as 3 percentage points could have a devastating effect on the predicted election outcome. (This whole area is largely ignored today, but the peril remains.)

He then compares the results of using polling predictions versus simple persistence predictions, and concludes that polling procedures at that time were not much more effective than persistence forecasts, at least during the four elections for which he made the comparison.

The conclusions of Chapter V serve as a pointed and relevant summary (p. 80):

1. The errors in 1948 were not much out of line with those of previous elections, but this finding should not reduce interest in searching out and correcting basic sources of errors.

2. The polls have not yet been able to demonstrate superiority over persistence prediction under the conditions of the last twelve years.

3. Considering the present practices in polling and in forecasting from polls, there is no reason to believe that errors in magnitude such as those occurring in 1948 are unlikely to occur in future elections.

4. Basic research and improvement of methods will be needed in every part of the polling process before polling organizations will be able to make election forecasts with confidence.

5. Polling organizations have not adequately educated the public about the magnitude of their errors, and statements about the errors, instead of being regarded as useful information, are regarded by the public as hedges.

6. The polls have had more credit than was their due for their past election predictions, and they are now reaping an equally unjust amount of criticism for much the same errors that have occurred in the past.

The results of the study of the pre-election polls of 1948 were the product of a very short, intense period of activity. Not only were they scientifically sound, but they were also intelligible, even to laymen, who were perhaps the most important consumer, and to the press, who would disseminate the report's message.

The report served as a focus for discussion and as a summary of what was good and where improvements could be wrought. Political polling is now a staple of our lives, not only in the United States but in many other countries. The pre-election poll has been joined by the election day voter poll. The conduct and interpretation of both types of polls still involve a large element of art as well as science. The fact that the public still is insensitive to polling error was revealed by the "surprise" Reagan victory in 1980. However, it is safe to say today that polling is an integral part of our lives, although perhaps one of the less comfortable parts.

Polling today, by and large, has a history of acceptance and is perceived by the public to be accurate. This perception has grown out of a few decades of good results. Perhaps this favorable record can be credited in part to the good work that Fred Mosteller led so well in 1948. It went a long way toward quenching the fires of controversy that raged around the 1948 pre-election polls.

Statistical Problems of the Kinsey Report

William G. Cochran, Frederick Mosteller, and John W. Tukey. *Statistical Problems of the Kinsey Report.* Washington, D.C.: The American Statistical Association, 1954.

by
James A. Davis

When I think about Fred Mosteller, what leaps to mind is not the steamy problems of Sex but the unsexy problems of steam heat and pedagogy. It all goes back to 1951, when I was a first-year graduate student in sociology. It came to pass that a call went out for teaching fellows to assist Fred in the Sisyphean task of teaching elementary statistics in the old Social Relations department. The major requirement for the post was—sensibly enough—prior completion of an introductory statistics course. I was one of two graduate students who had performed that feat. Indeed, I had performed it twice: first as an undergraduate at Northwestern, where we concentrated on drawing histograms, and then at Wisconsin, where, as a first-year graduate student, I listened to a famous sociologist lecture to a class of 200 in a very low voice, at 8 a.m., while continually standing sideways and gazing out the window. In other words, when I was called to the Harvard faculty, I knew about as much statistics as a stone.

No problem. With the help of the Hagood text[3] and Fred's lectures, I learned statistics one jump ahead of the troops. For teachers there are few greater incentives. And I learned the two p's of statistics teaching and consulting—patience and practicality.

Lord, he was patient. I shall never forget his filling not one but two blackboards in Emerson Hall with the inscrutable minutiae of ANOVA, only to receive a classic dumb question that indicated the student had become lost at the first turn. Fred smiled his "Gee, golly, shucks" smile, erased both blackboards, and started again from the beginning. That's patience.

As for practicality, the assistants shared office space with Fred, so I got to eavesdrop on his frequent unpaid consultancies. In fact, I learned how to do it myself. It appeared that the task of the consultant, when presented with two elaborate alternatives for laborious work, neither of which quite met the assumptions of the procedure, is to say, "Uh huh, yes—now, which one would you rather do? Uh huh, that would be fine—but you might want to read up a little on ...." At the time, I wondered whether this was really what was meant by "high-powered statistical consulting." I now know it

[3]Margaret Jarman Hagood and Daniel O. Price. *Statistics for Sociologists*, revised edition. New York: Holt, 1952.

is—the high-powered consultant helps the client to find legitimate ways to do what he or she really wants; the low-powered consultant beats them over the head about assumptions.

But that's not the ultimate on practicality. One night I was sitting at my desk grading exams when the great man ambled over, blinked, and said, "I understand you are interested in becoming a teacher." "Yes, sir," I said. "Do you want to know the key to success in a classroom?" "Oh, yes." "Every time you go into a classroom, open all the windows.. [two beat pause] ..keeps them awake." Perhaps it is an apt commentary on the efficacy of consulting that I now teach in a glass box where none of the windows can be opened and the air is turned off periodically to save money.

Now, the sex part. In 1954 Bill Cochran, Fred, and John Tukey published a book-length critique of the methodology in Kinsey, Pomeroy, and Martin's 1948 book, *Sexual Behavior in the Human Male.* Nowadays it is hard to grasp the dither surrounding the book, but in the early 1950s it was exceedingly controversial. The problem was that Kinsey was a statistical Grandma Moses, talented but untrained and as innocent of the concept of statistical inference as I had been when I finished my two intro stat courses. Consequently, his sampling was unprofessional, and his confusing justifications of it made things worse. But Kinsey was light years ahead of anyone else in the sex game scientifically, and a harsh critique would have set back an important line of behavioral research.

Given the 1950s belief that sex surveys of random samples were impossible (disproved soon after by NORC, but who could have known that?), the committee tactfully pointed out some lapses (it appears Kinsey was shaky on the definition of medians) and basically took the Mostellerish position that, considering the practical options open to him, he should keep doing what he was doing, but perhaps read up a bit on .... The book is laden with pithy tart aphorisms showing extraordinary even-handedness toward Kinsey and toward his critics. Among my favorites: "To summarize in another way ... (i) they did not shirk hard work and (ii) their summaries were shrewd descriptive statements rather than inferential statements about clearly defined populations." Or, on interviewing: "The interviewing technique has been subjected to many criticisms but on examination the criticisms amount to saying 'answer is unknown' .... These conclusions can be summarized by saying that we need to know more about interviewing in general."

To speak clearly when you know the answer (e.g., open windows), to push the client gently toward the light (e.g., Kinsey's sampling), and to shut up when you know no one knows the answer (e.g., interviewing), i.e., to be patient and practical: these are the marks of a master craftsman.

Stochastic Models for Learning

Robert R. Bush and Frederick Mosteller. *Stochastic Models for Learning.* New York: Wiley, London: Chapman and Hall, 1955.

by
Paul W. Holland[4]

"... A friendly challenge from an adventuring physicist and a mathematical statistician." So a sympathetic reviewer for *Contemporary Psychology* described *Stochastic Models for Learning* a year after its publication. The "adventuring physicist" was the late Robert R. Bush, whose subsequent distinguished career included the chairmanship of the Psychology Department at the University of Pennsylvania. The mathematical statistician is, of course, our friend and colleague, honored by this volume—Frederick Mosteller.

The publication of *Stochastic Models for Learning* in 1955 was instrumental in defining and launching the merger of theoretical and experimental psychology that became known as "mathematical psychology." Certainly by 1962, when I entered graduate school, "Bush and Mosteller," as it was called, was a standard reference and text for those of my fellow students at Stanford who studied this relatively new field. It gave the first book-length mathematical and statistical analysis relevant, in great detail, to real psychological experiments involving "learning," broadly conceived. It treated a class of models that could be applied to a wide variety of problems, *and it applied them to real data.* These models could often provide simple, tractable, and powerful tools for analyzing data from complex learning experiments. As a math major in search of a bridge to the behavioral sciences, "Bush and Mosteller" was a revelation to me. Here was a book that used real mathematics in real problems that involved detailed human behavior!

The models described in *Stochastic Models for Learning* are elegantly simple in conception. In the simplest case, a sequence of "learning trials" is hypothesized, and on the nth trial the learning organism can make one of two responses—say A or B. This behavior is viewed as stochastically determined by a response probability vector

$$p_n = (p_{An}, p_{Bn}), \quad p_{An} + p_{Bn} = 1.$$

The response probability vector for the next trial, p_{n+1}, is determined both by the value of p_n and by some outcome or event that occurs on the nth trial (exactly what these outcomes or events *are* depends on the details

[4] I would like to thank Professors William K. Estes and M. Frank Norman, as well as the staff of the ETS Library, for their generous help in the preparation of this review.

of the learning experiment). Bush and Mosteller assume there are t such events, $E_1, \ldots, E_t$, only one of which can occur on each trial. If E_j occurs on trial n, then p_{n+1} is given by

$$p_{n+1} = T_j p_n$$

where T_j is a matrix operator and maps probability vectors into new probability vectors. The models of Bush and Mosteller are also called "linear models" and "operator models." Clearly, the restriction to two possible responses of the organism on each trial is not essential.

Their abstract framework does not specify what the "trials" are, what the possible responses might be, or what the "events" are. These depend on the details of each specific experiment. The coefficients of the operators T_j are the "learning" parameters of the model and are to be estimated from data. The interpretation of these learning parameters also depends on the structure of the experiment to which they are applied. Estimates of these learning parameters were Bush and Mosteller's answer to the crude averages and other summary statistics that their forerunners had used to summarize data from learning experiments.

"Bush and Mosteller" applies these general models to such diverse topics as free-recall verbal learning, avoidance training of dogs, social imitation in children, rats running in T-mazes, and symmetric choice-prediction in humans (which is a sort of T-maze for people). Certainly, Bush and Mosteller's broad application of the operator models was a major factor in their early popularity, and was a significant facet of the initial development of the field of mathematical psychology itself.

As a book on the application of mathematics and statistics to a scientific problem, *Stochastic Models* is a model well worth emulating. The first half contains detailed mathematical and statistical derivations for the general models, and the second half describes many applications of the general results. Bush and Mosteller include a healthy dose of the necessary mathematical prerequisites, because their intended audience is "the experimental psychologist with only a limited background in mathematics." Each chapter ends with a summary and references, and the book is extremely well written. However, in no sense is "Bush and Mosteller" a cookbook. Principles are put forth for the reader to apply to new problems. Derivations are given that help the reader obtain relevant results for new experiments. However, psychologists were not the only audience. Probabilists found novel stochastic processes to analyze (Iosifescu and Theodorescu, 1969), and statisticians found difficult estimation and testing problems lurking behind these models.

Stochastic Models for Learning appeared in the midst of an era of active research on the mathematical and statistical analysis of learning experiments. Other books followed in the tradition of mathematical and statistical rigor and of psychological relevance that "Bush and Mosteller" helped to foster. Examples are Bush and Estes (1959), Arrow, Karlin and Suppes (1960), Luce, Bush and Galanter (1963), Atkinson (1964), Atkinson, Bower and Crothers (1965), Norman (1972), and Krantz, Atkinson, Luce and Suppes (1974).

Much of the material in *Stochastic Models for Learning* came directly from research articles by Bush, Mosteller, and their colleagues. However, other scientists were contributing to the development of mathematical psychology at about the same time. Notable among them was W.K. Estes, to whose work Bush and Mosteller devote one chapter. Estes had developed an alternative class of learning models—called stimulus sampling models. In a sense, Estes's models are more general than the operator models. For example, Estes and Suppes (1974) showed that in many cases operator models can be regarded as the limits of sequences of stimulus sampling models. More substantively, during the 1960s, operator models and stimulus sampling models were compared head-to-head in terms of their predictions for various types of human learning experiments, with the general finding that in many situations the predictions of the stimulus sampling models fared better than those of the operator models—see, for example, the discussion of Atkinson, Bower and Crothers (1965, page 116). As time went on, the nature of the mathematical models used in cognitive science evolved. Instead of the simple "response probability" of Bush and Mosteller, which changed over time with the experiences of the organism, modern models involve memory processes, learning states, and various other features that interact in complex ways. Current issues of the *Journal of Mathematical Psychology* (founded in 1964) illustrate these changes. However, the exciting mathematical, statistical, and psychological mixture initiated by "Bush and Mosteller" still appears in that and related journals.

And yet, despite all these advances and the increased complexity, new types of operator models continue to show up as appropriate and useful models in contemporary psychology. For example, a model of Pavlovian conditioning due to Rescorla and Wagner (1972) uses an operator model closely related to those of Bush and Mosteller. It is regarded as a very successful model of Pavlovian conditioning (Schwartz, 1978). Similarly, a model for discrimination learning developed by Zeaman and House (1963) and Lovejoy (1966) and analyzed by Norman (1974, 1976) is an operator model in which *two* probability vectors describe the state of learning of the organism on each trial. One describes the response probabilities, and the other the probability of attention of the organism to the relevant stimulus.

Finally, Estes (1984) recently found that a simple linear model gave a satisfactory explanation of the data in an experiment in which reinforcement probabilities varied in a sinusoidal pattern. The models of Bush and Mosteller may be 30 years old, but they are still in use! Estes's (1956) prediction that these models would not go away has proved correct.

The first page of the preface of *Stochastic Models for Learning* states that:

> The system we describe in rather general terms is applied to a number of particular experimental problems, but we make no pretense at completeness or finality and shall feel much rewarded if we provide a start on a good approach.

After thirty years, *Stochastic Models for Learning* is still cited by authors in a variety of fields, including engineering and management science, and well in excess of typical citation rates (as a check of the Science Citation Index reveals); the models developed there are still in use to the point of being the common knowledge of those who study cognitive science; and generations of students have followed its clear derivations and examples. I would say that the authors of *Stochastic Models for Learning* ought to have many reasons to feel "much rewarded."

References

Arrow, K., Karlin, S., and Suppes, P., Eds. (1960). *Mathematical Methods in the Social Sciences.* Stanford, CA: Stanford University Press.

Atkinson, R., Ed. (1964). *Studies in Mathematical Psychology.* Stanford, CA: Stanford University Press.

Atkinson, R., Bower, G., and Crothers, E. (1965). *An Introduction to Mathematical Learning Theory.* New York: Wiley.

Bush, R. and Estes, W., Eds. (1959). *Studies in Mathematical Learning Theory.* Stanford, CA: Stanford University Press.

Estes, W. (1956). Review of *Stochastic Models for Learning. Contemporary Psychology*, **1**, 99–101.

Estes, W. (1984). Global and local control of choice behavior by cyclically varying outcome probabilities. *Journal of Experimental Psychology*, **10**, 258–270.

Estes, W. and Suppes, P. Foundations of statistical learning theory, II. The stimulus sampling model. In *Contemporary Developments in Mathematical Psychology, Volume I*, edited by D.H. Krantz et al. San Francisco: W.H. Freeman and Company, 1974. pp. 163–184.

Iosifescu, M. and Theodorescu, R. (1969). *Random Processes and Learning.* New York: Springer-Verlag.

Krantz, D., Atkinson, R., Luce, D., and Suppes, P., Eds. (1974). *Contemporary Developments in Mathematical Psychology.* San Francisco: W.H. Freeman and Company.

Lovejoy, E. (1966). Analysis of the overlearning reversal effect. *Psychological Review*, **73**, 87–103.

Luce, D., Bush, R., and Galanter, G., Eds. (1963). *Handbook of Mathematical Psychology, Volumes I and II.* New York: Wiley.

Norman, M.F. (1972). *Markov Processes and Learning Models.* New York: Academic Press.

Norman, M.F. Effects of overtraining, problem shifts, and probabilistic reinforcement in discrimination learning: Predictions of an attentional model. In *Contemporary Developments in Mathematical Psychology, Volume I*, edited by D.H. Krantz et al. San Francisco: W.H. Freeman and Company, 1974. pp. 185–208.

Norman, M.F. (1976). Optional shift, and discrimination learning with redundant relevant dimensions: Predictions of an attentional model. *Journal of Mathematical Psychology*, **14**, 130–143.

Rescorla, R. and Wagner, A. A theory of Pavlovian conditioning: Variations in the effectiveness of reinforcement and non-reinforcement. In *Classical Conditioning* II, edited by A.H. Black and W.F. Prokosy. New York: Appleton-Century-Crofts, 1972.

Schwartz, B. (1978). *Psychology of Learning and Behavior.* New York: W.W. Norton and Company.

Zeaman, D. and House, B. The role of attention in retardate discrimination learning. In *Handbook of Mental Deficiency*, edited by N.R. Ellis. New York: McGraw-Hill, 1963.

Probability with Statistical Applications

Frederick Mosteller, Robert E.K. Rourke, and George B. Thomas, Jr. *Probability with Statistical Applications.* Reading, MA: Addison-Wesley, 1961.

Probability with Statistical Applications. Second Edition. Reading, MA: Addison-Wesley, 1970.

by
Emanuel Parzen

Fred Mosteller's professional contributions have benefited and influenced almost all contemporary statisticians, as well as an extremely large number of users of statistical techniques. One pervasive feature of Fred's contributions is their continuous relevance. This aspect became clear to me when I undertook to review from the perspective of the 1980s the book *Probability with Statistical Applications* (PWSA), published in 1961 and, in a revised second edition, in 1970.

My first contact with Fred's many outstanding contributions to college education in the United States came in 1947–1948, as his student in Math 9 during my junior year at Harvard. My field of concentration was mathematics, and Math 9 was a year-long first course in probability and statistics. Richard von Mises, who had been teaching this course, had just retired. To us math majors it was a mystery which professors in the Harvard Mathematics Department were qualified to teach it. As it happened, two shared it: Lynn Loomis (an expert in measure theory) taught probability the first semester, and Fred Mosteller taught mathematical statistics the second semester. That course is indelibly imprinted on my mind because Chebyshev's inequality made its first appearance in the course as a problem on an exam. As a teacher, I have never attempted to introduce Chebyshev's inequality as an exam problem, and I have often pointed out to my classes that my failure to do so must prove something (perhaps that I'm not providing as good an education as I might).

PWSA was written as a textbook that would be accessible to students who have completed two years of high school algebra. When one compares PWSA with the many books for introductory statistics courses that are constantly being published, one sees that the new books (which are often similar to each other) bear little resemblance to PWSA. In the "Era of Sputnik," when PWSA first appeared, national interests were thought to require that students of all levels be taught the glory and applicability of probability theory. This broad effort included a televised course that Fred Mosteller taught in 1961 as part of "Continental Classroom," a daily educational program on the NBC television network. *Probability and Statistics,* the text for that course, was derived from PWSA by abridgment and slight rewriting.

It is important to re-read PWSA, because it is an excellent book and because we need to be reminded of the topics taught to hundreds of thousands of students in a United States eager for excellence. In the 1960s the challenge was to keep up with the Russian space program; in the 1980s the challenge seems to be to keep up with the Japanese trade program. Are the topics in PWSA no longer vital as part of introductory courses? Should we be concerned to ask where they have gone in our educational programs?

Some of the distinctive features of PWSA are: an elementary introduction to random variables and their distributions; an intuitive introduction to continuous random variables and the normal distribution, which lays the foundation for the theory of sampling routinely used in statistical inference; a new development of the central limit theorem, introduced by approximating binomial probabilities; a treatment of modern Bayesian inference; a treatment of least squares that depends only on high school algebra; an extensive set of applications of statistical inference presented in a unified way.

An aficionado of elementary books would discover other distinctive features as well. I like the table (p. 336 of first edition) of "analogous theorems for triangles and for random variables." I love Table VII, inverse (quantile function) of the cumulative normal distribution. The quantile function certainly belongs in a modern introductory course.

The most important comparison of PWSA with current introductory statistics texts is that two-thirds of PWSA is an introduction to probability and one-third is an introduction to statistics. Many current introductory statistics books omit probability and feature methods of exploratory data analysis.

PWSA skillfully blends mathematical reasoning, practical thinking, interesting illustrative examples, and challenging exercises. It does not provide a cookbook survey of statistical methods, to which students are exposed in many introductory courses in statistics. Rather it treats topics that represent the *foundation* on which understanding is built. PWSA is valuable historically as a landmark of the intellectual climate in which it was written. In my judgment it provides a valuable benchmark against which to measure courses currently being taught. All teachers of introductory statistics courses and authors of introductory statistics textbooks should read PWSA occasionally to be sure that they are giving their students access to its insights.

Inference and Disputed Authorship

Frederick Mosteller and David L. Wallace. *Inference and Disputed Authorship: The Federalist.* Reading, MA: Addison-Wesley, 1964.

Applied Bayesian and Classical Inference: The Case of the Federalist Papers. New York: Springer-Verlag, 1984.

by
John W. Pratt

This exemplary study developed and applied pioneering methodology to resolve a daunting problem in a field where even rudimentary statistical methods had been little used: distinguishing authorship. The problem is not quantitative on its face, or readily quantified. Weak data doomed simple solutions, and the task eventually taken on was to differentiate two writers solely by their rates of use of ordinary English words.

Mining such low-grade ore was expensive with the technology of the time and called for care and judgment in a strategy of successive sifting. Text was processed in waves and unpromising words discarded. Even so, the initial digging required substantial resources, not easy to marshal and manage. Refining the ore posed major methodological problems, solved by a number of approaches, some quite deep. A complex model was developed, checked, and used. Formidable calculations were approximated. All the work was impeccably done and documented. The analysis is as close to ideal as one can imagine. Every conceivable concern seems to be anticipated and conscientiously addressed, often by extensive and clever additional analyses. The exposition takes great pains to be helpful to all types of reader. In every way, the work surpasses anything one would ordinarily think of asking for. Such perfection renders review so superfluous that I resisted the assignment, but it does raise some discussable issues. First, though, I will describe the instructive main methodology, an empirical Bayes analysis of a negative-binomial, random-parameter model that disentangles regression, selection, and shrinkage in estimation and discrimination.

Background

Of the 85 pseudonymous *Federalist* papers written to persuade New York State citizens to ratify the U.S. Constitution, 51 are known to be by Alexander Hamilton, 14 by James Madison, 5 by John Jay, and 3 by Madison with input from Hamilton. The question is who wrote the other 12; historians favor Madison, but Hamilton is also possible.

In earlier statistical work, Frederick Williams and Frederick Mosteller found that the length of Hamilton's sentences in undisputed *Federalist* papers had mean 34.55 words and standard deviation 19.2 words, versus Madison's 34.59 and 20.3. This suggests the difficulty of differentiation (and of reading stylish writing of 200 years ago). Williams and Mosteller also carried out a discriminant analysis based on the frequency of one- and two-letter words, nouns, and adjectives. This supported "the convictions of modern linguists that the categories of Latin grammar are often ill fitted to describe English" (p. 9) and indicated who wrote the disputed papers if all were known to have the same author, but didn't classify them individually with any assurance. About 1959, Mosteller learned from a historian of two marker words: where Hamilton used "while," Madison used "whilst." Mosteller and Wallace then launched an investigation based on word counts. In a pilot study they found two more markers: in known papers, Madison used "upon" rarely and "enough" never, but Hamilton used them (together) between 2 and 13 times per paper. The disputed and joint papers are quite strongly discriminated by these words, but there are clearly statistical, editing, typesetting, and other dangers in relying on just four low-frequency words.

The Methodology of the Main Study

Introduction

We—Mosteller and Wallace, and now you and I—seek the likelihood ratio for a paper of unknown authorship, the probability of the data if the paper is by Hamilton divided by the same probability for Madison. Multiplying this likelihood ratio by your prior odds for Hamilton's authorship would give your posterior odds, by Bayes's theorem (p. 54), but we won't do this because it depends on you, is trivial to do, and adds nothing in view of the overwhelming likelihood ratio we ultimately obtain.

Oversimplifying to get started, suppose all we know about an unknown paper is that the word "kind" occurs x times in w thousand words. Model Hamilton's use of the word "kind" by a Poisson process with rate μ_H per thousand words and Madison's similarly with rate μ_M. If μ_H and μ_M are known, the likelihood ratio is the probability of x for a Poisson distribution with mean $w\mu_H$, say $P(x \mid w\mu_H)$, divided by the probability for mean $w\mu_M$ (p. 55).

If μ_H is unknown but has a known distribution, and if the unknown paper is by Hamilton, then the probability of x is the marginal probability obtained from the conditional probability $P(x \mid w\mu_H)$ and the known (marginal) distribution of μ_H, the integral or expectation of the former with respect to the latter. If the distribution of μ_H is viewed as a prior distribution, be it "subjective" or, as here, data-based, then the marginal distribution of x is a "predictive" or "forecast" or "preposterior" distribution. If μ_M also has a known distribution, then regardless of terminology, the likelihood ratio is simply the marginal probability of x given Hamilton's authorship divided by the marginal probability given Madison's authorship. Would that the calculation were as simple as the concept, especially for the more realistic model we shall come to!

If we have information on several words, the likelihood ratio in general is the ratio of joint probabilities of the data given authorship. However, if we model each word as before and in addition take occurrences to be independent across words, then the likelihood ratio for the set of words is the product of the likelihood ratios for each word. Specifically, for each author, assume that the occurrence frequencies of different words are conditionally independent given the rate parameters and that the rate parameters are marginally independent. Then the occurrence frequencies are marginally independent, the joint probability of the frequencies is the product of their individual probabilities, and hence the joint likelihood ratio is the product of the individual likelihood ratios.

We use *Federalist* and other papers known to be by Hamilton and Madison to learn about the rates and their distributions and the inadequacies of the foregoing model. For statistical analysis we need some kind of stability, such as constant parameters, across papers. Context-dependent words would be very hard to model in any stable way and would provide weak statistical evidence anyhow, so we omit them. Even for others, we expect the Poisson distribution with constant rates to be inadequate because of local contagion and avoidance, and the data show more-than-Poisson variation in occurrences of individual words but reasonable accord with the negative binomial distribution (pp. 32–33), which we therefore use. Some dependence across words is inevitable, but the impossibility of simultaneous occurrence of two words contributes very little dependence even at the highest frequencies, and the data indicate that most correlations between words are small (p. 36). Since dealing with dependence is dicey indeed, we defer it to a post-analysis adjustment.

The Negative Binomial Model with Random Parameters

We arrive thus at a model with four negative binomial parameters for each word (two for each author) and independence across words and papers (given the parameters). We incorporate paper length without additional parameters, much as for the Poisson (see below). If the parameters, or their distributions, were known or accurately estimated, we could obtain the likelihood ratio for an unknown paper as described above for the Poisson model. Unfortunately, we have too little information about too many parameters to proceed in any simple or simple-minded way, Bayesian or not, and one more important step is needed. Note that, a priori, one non-contextual word is much like another for our purposes. Accordingly we develop a random-parameter model akin to Model II ANOVA, with a few adjustable constants, called "hyperparameters" nowadays ("underlying constants" in the book).

Given the hyperparameters, the model specifies the distribution of the negative binomial parameters for each word, with independence across words. We may consider any such distribution a prior distribution. A purely (or puerilely) Bayesian analysis would introduce a (hyper?) prior distribution on the hyperparameters, but we will do sensitivity analysis instead. This will be far more convincing, since the main conclusion holds for all empirically plausible hyperparameter values and more. Indeed, no naughty "subjective" distribution is ever needed, and the analysis, though empirical Bayes, has such a strong empirical base that it should pass frequentist muster.

Recapitulation

Deferring many details but adding some notation, we now sketch the whole model (Figure 3.1–1 and Section 4.3). Let β denote the hyperparameter vector and γ denote a parameterization of two negative binomial distributions. Each β determines a distribution (prior to Stage I) from which values of γ are drawn independently, one for each word. Each value of γ determines two negative binomial parameters for each author. For each paper by each author, a frequency is drawn for each word from the negative binomial distribution determined by that word's value of γ, the author, and the paper length, independently across papers and words.

In Stage I of the analysis, word frequencies $\mathbf{X}$ are counted in a set of papers of known authorship. Given $\mathbf{X}$ (and β), the conditional distribution of the values of γ for all words is determined by Bayes's theorem. In this distribution, which is posterior to Stage I but prior to Stage II, the values of γ are still independent across words.

In Stage II of the analysis, word frequencies $\mathbf{Y}$ are observed in a paper of unknown authorship. For each author, the probability of $\mathbf{Y}$ given $\mathbf{X}$ is determined. It is the marginal probability resulting from the distribution of negative binomial parameters for that author determined in Stage I and independent negative binomial word frequencies given the parameters, and it still evinces independence across words. The likelihood ratio is the ratio of the probabilities of $\mathbf{Y}$ given $\mathbf{X}$ for the two authors.

Remaining Details of the Model

The following representation of the negative binomial distribution as a mixture also motivates a particular way of incorporating paper length (Section 4.1). Suppose that, for each author, each word has a Poisson rate in each paper that varies from paper to paper like independent drawings from a gamma distribution. If the gamma distribution has gamma parameter κ and scale δ, and hence mean $\mu = \kappa\delta$ and variance $\mu\delta$, then in a paper of length w, the word's frequency x is negative binomial with mass function

$\binom{x+\kappa-1}{x}(w\delta)^x/(1+w\delta)^{x+\kappa}$, mean $w\mu$, and variance $w\mu(1+w\delta)$. This approaches the Poisson distribution as $\delta \to 0$ with μ fixed ($\kappa = \mu/\delta \to \infty$). Thus δ is a measure of non-Poissonness. It can be interpreted as the excess variance relative to the Poisson per thousand words. This model fits the data well, whereas the Poisson tails are clearly too low (pp. 32–35).

We still need to develop a random-parameter model for the variation across words of the four negative binomial parameters (two per author) belonging to each word. We reparameterize so that a model with a few adjustable hyperparameters β_i can describe this variation simply yet adequately for our purpose (Section 3.2). The sum of the rates $\sigma = \mu_H + \mu_M$ can be estimated fairly well for each word from that word's data alone, so treating it as uniformly distributed suffices. What really matters is the distribution of the other parameters given σ, where the following model fits the data adequately. The distribution of $\tau = \mu_H/\sigma$ given σ is symmetric beta with parameter $\beta_1 + \beta_2\sigma$. The non-Poissonnesses δ_H and δ_M are independent of the rates. After transformation to $\zeta = \log(1+\delta)$, the sum $\xi = \zeta_H + \zeta_M$ is gamma distributed with mean β_4 and gamma parameter β_5, and $\eta = \zeta_H/\xi$ is independently symmetrically beta distributed with parameter β_3. This model has five adjustable hyperparameters β_i.

Selecting Discriminators

The selection of words to count called for great care, because counting was expensive and time-consuming, especially with that era's technology (Section 2.5). Also, a plethora of weakly discriminating words might magnify defects in the model and computational approximations and add noise all around. However, we can decide to drop words from further consideration at any time before observing their frequencies in unknown papers without affecting the analysis of the remaining words, provided our random-parameter model describes the way the words retained were originally drawn (pp. 51–52, 67–68). The empirical-Bayes shrinkage estimates the regression effect. (All variables regress. The real issue is not selection but flat priors when data are relatively weak or when errors accumulate systematically over many variables. Similar comments apply to non-Bayesian analyses.) It is important that the words used to estimate the random-parameter model be representative of the original pool of possible words as regards behavior given frequency. They were carefully chosen to be so, although weighted toward high frequency because high-frequency words are fewer but relatively important (pp. 67–68).

What Else Did They Do?

Are you exhausted? I am, and I haven't begun to tell or even mention it all. Bear in mind that the model is far more complex than, for instance, unbalanced Model II ANOVA. They extract from the data estimates of the hyperparameters β of the random-parameter model and limits on their plausible ranges (Section 4.5). They approximate the four-dimensional integrals needed for the marginal probability for each word and author in Stage II by treating the Stage I posterior (Stage II prior) as concentrated at its mode. They study the errors in this approximation by a combination of transformation to approximate normality, the delta method, and the Laplace integral expansion (Section 4.6). They study the effect of correlation between words (Section 4.7). They compare the actual and theoretically expected performance of the model on the unknown papers, the papers used in the estimation, and four other Hamilton papers—for individual words, groups, and the whole final set—and find that the model is adequate, in particular to estimate regression effects (Sections 3.4–3.6 and 4.8). They compare the Poisson and negative binomial results and find that using the inappropriate Poisson model can easily square the odds, increasing the log odds by 50–150% (p. 75). They search out and eliminate words whose Madison rate varied over the 25-year period of Madison writing they were using to supplement the relatively few known Madison *Federalist* papers (Sections 2.2 and 4.10). They develop some pieces of new theory. They discuss practical problems of counting words, the separation of historical from statistical evidence, possible outrageous events and the greater impact of their possibility on strong than weak evidence, and many more such matters.

Beyond the main study, they present three complete, self-contained analyses by other methods (Chapters 6–8) and a review of the state of statistical authorship studies at the time of the second edition (Chapter 10).

The Ideal and the Real

Isn't it wonderful to see such high standards set and met in a real study of a real problem? Yes indeed, but its very matchlessness prompts some musings on statistical publication and statistics in our society.

First, how should the theoretical and methodological literature reflect applications? Are too many minor advances extracted from real or mythical problems, sanitized, rigorized, polished, and presented in pristine mathematical form and multiple forums, in senseless generality or least publishable units? If so, how much is due to the field's maturity, its growth, the dependence of reputation on media blitz, funding and academic promotion procedures, hard criteria driving out more important softer ones? Would it help if theoreticians did more serious applied statistics, as is often urged?

What do genuine statisticians do? Mosteller and Wallace first published "Notes" in a symposium proceedings, then a summary account in *JASA*, and finally a 267-page book. Had time allowed them to repackage their many contributions separately with their usual care, the benefits in accessibility and recognition by later comers would have compensated for any duplication.

Theoretical and methodological essences distilled under present journal pressures inevitably smell somewhat artificial, but this doesn't offend me greatly. Though I can't begin to take everything in, still less could I without distillation. I do dislike excess repetition and least publishable units, still more the shameless claims and self-promoting styles apparently deemed critical to esteem in some circles. Perhaps few are fooled and I am foolish to fulminate. But I wish I more often encountered attitudes and statements like "We feel obliged to provide treatments of techniques that are not new with us and are known to many practicing statisticians, yet, as far as we know, are not available in the literature." (p. 92)

Even if current publication practices are annoying, redundant, unfair, and misleading, widespread recognition of this mitigates some of its effects, and prospects for improvement are limited by human nature. But contemplating this book also brings to my mind a much more serious set of problems that we are facing up to much less.

The *Federalist* study represents an ideal. Two outstanding principal investigators and an army of over 80 Harvard students and others labored painstakingly over several years. No trouble was spared. All bases were uncovered and covered. Objective criteria for statistical procedures were spelled out in advance as far as humanly possible. Selection effects were fully reported and allowed for. Choices were made without knowing which side they would favor (unusually feasible here, and the investigators unusually scrupulous). And, as it happens, the outcome was unequivocal. We are compelled to believe it.

But such efforts are seldom feasible or justified. How should we commission, manage, and react to more everyday studies? Why do we believe this one, really? Because an army covered every base? Because we followed every step of the argument and it convinced us?? No—because we saw that it convinced the authors and we trust them. And had the evidence been equivocal, we would be relying even more on their honesty, competence, judgment, and even-handedness.

If the *Federalist* dispute were being aired in court or in the media and this study were presented and paid for by the side it favors, how would it fare? I think the other side's statisticians and advocates would have no trouble raising dust by such time-honored methods as ridiculing assumptions taken literally (Section 4.3) and finding no statistically significant difference by a weaker but more robust test (perhaps simpler and more credible to a court or the public), or by less honorable means. And if the issue could be clouded here, what about an analysis of more equivocal evidence by more peccable people? "In Britain, Morton [a leading stylometrician] finds the path of the expert literary witness a very rocky road .... Opposing counsel treat matters that would make no difference in scientific investigations as if they are deliberate attempts at fraud. He found the treatment of the expert witness in this new field repelling ..." (p. 278). Are manners or matters any better in the U.S.?

Many controversial public and legal issues call for statistical analyses. All statisticians must often cringe at the unscrupulous way columnists, editorializers, and proselytizers use statistics—*that's* what's criminal! Furthermore, analyses by interested parties, however well-meaning (even statisticians), do not give a clear and balanced picture, yet usually drown out disinterested analyses if there are any. We are not confronting the impracticality of adversarial approaches to resolving applied statistical questions. It's not just the expense, but the impossibility of seeing where the balance lies. Licensing statisticians or setting statistical standards wouldn't help much. Expert systems won't help. Even if a tell-all computer program records every regression tried in the rough-and-tumble of analyzing observational data, will we really report every step with its rationale? Would anyone read it or believe it if we did?

We must recognize that credible analyses need not be expensive and expensive analyses need not be credible. The truth is rarely as one-sided as here. Resources are rarely as great. Analysts and analyses almost never remotely approach the calibre, thoroughness, and integrity of those here. What could be approached more often is their neutrality and even-handedness, but this would require radical change in our ways of doing business, especially in the media and the courts. The problem extends to many fields, but is probably particularly severe in statistics by the nature of statistical evidence. National Research Council studies and the like commandeer top experts for no pay and are still expensive—two reasons they can't be standard procedure. Yet neutral auspices are essential to balancing complex evidence.

How can we teach people to ignore selective statistics and "examples" and their selectors? The truth prevails eventually in science, but in other arenas, playing out controversies as we do now is too slow and costly to yield good decisions or even justice. How can a democracy arrive at effective procedures for using statistics? Fred Mosteller's wisdom on these issues would be valued by all who know him.

Postscripts

1. Whodunit? James Madison—at conservatively adjusted odds of 33:1 for one paper, 180:1 for another, and beyond a shadow of a doubt (2500:1) for all others. But the deduction (really induction) was not elementary, my dear Watson!

2. I found it helpful to relate regression and selection effects and shrinkage in estimation and discrimination in the simplest normal-theory case to the Mosteller-Wallace main anal-

ysis. Notes available on request.

3. Though perhaps no easier to read, this review is easy to distinguish from stylish writing of Revolutionary times: the length of sentences in nontechnical sections has mean 17.56 words and standard deviation 10.4 words (defined and counted by Word Perfect 5.0 cursor commands).

4. Further sources of recent references include Holmes (1985).

5. I promise to resist any temptation to publish any of this again.

Reference

Holmes, D.I. (1985). The analysis of literary style—a review. *Journal of the Royal Statistical Society, Series A*, **148**, 328–341.

Fifty Challenging Problems in Probability

Frederick Mosteller. *Fifty Challenging Problems in Probability with Solutions.* Reading, MA: Addison-Wesley, 1965.

by
Joseph I. Naus

"Have I got a problem! My son Marvin never visits me. Each night I put up a roast, bake his favorite cake. Not like it's a big trip from 72nd Street where he works to me at 92nd Street. One stop on the IRT Express. He's a good boy, Marvin, but you know how it is at rush hour; everybody rushes to the platform where the train is. It's push, push, push. Before he knows it, Marvin is on the downtown IRT. What else can he do? They pack you in the trains like sardines. By the time he can get off, he's at Flatbush Avenue in Brooklyn, near his girlfriend Sue's house; you know, the one who can't cook. He says there are as many uptown as downtown trains, so it should even out. I'm still waiting. When once in a blue moon the train to me comes first, I see how thin my poor Marvin is."

"You've got a problem! At least your son is seeing someone. My son Sheldon was going with such a nice girl. A lovely girl, and he just stops seeing her. 'Not right for me, Mom,' he says. I tell him, 'You're not so young anymore, Sheldon; at forty-five it's time you settle down. Every year since your twenty-seventh birthday it's been a different girl you've been going steady with—eighteen lovely girls. What are you waiting for, Sheldon, a dream?' He says, 'Don't worry, Mom, I have it all planned out. If my new date tonight is nicer than the others, then I'll marry her.' He says it's an idea he read about when he was twenty-seven, about how to have the best chance to marry the nicest girl. I ask him, 'Where did you read this, Sheldon, in Ann Landers?' He says, 'No, Fred Mosteller.' "

"Do you think this Fred Mosteller can help me with my son Marvin?"

"Why not? My son Sheldon says that Fred Mosteller gives great solutions to many challenging problems."

When he wrote *Fifty Challenging Problems in Probability*, Frederick Mosteller gave the solutions to fifty-six problems. (Fred always gives more than promised.) Among them are probability problems that lead naturally to famous questions in number theory (Problem 1: The Sock Drawer, and Problem 56: Molina's Urns), as well as a variety of questions that add intuition and interpretation to quantities such as π and e. The solutions to the problems develop a variety of approaches, including induction (Problem 17: Twin Knights), indifference arguments (Problem 47: Choosing the Largest Dowry), building from smaller examples to general solutions, geometric approaches, combinatorial techniques, symmetry arguments (Problem 22: The Ballot Box), using inequalities, maximizing, approximations and limits, and direct attack. The book also forces the reader to look carefully into basic concepts, such as what is meant by "at random" (Problem 25: Lengths of Random Chords), and what is meant by a "sample space" (Problem 13: The Prisoner's Dilemma).

These ideas, techniques, and approaches yield a group of problems that are elegantly simple in their statement. I have used various of the problems with classes of students who were not at all keen on mathematics. I have also used the problems with honors undergraduates and graduate students. They all understood immediately questions like "The Prisoner's Dilemma." (The solution took somewhat longer for them to understand.) Similarly, Problem 49: Doubling Your Accuracy, and Problem 21: Should You Sample with or without Replacement, were very useful in teaching students not to take things for granted.

The problems in *Fifty Challenging Problems in Probability* are fun, demanding, and instructive; but, more than that, they introduce students to a variety of research areas. Two examples are the Birthday Problem and the Secretary Problem. Problems 31 to 34 deal with various interpretations of the Birthday Problem. These problems generalize readily to more difficult problems. Fred pushes in this direction at the end of Problem 31; the reader is asked to generalize the classical Birthday Problem to find "the least number to achieve at least one pair of either identical birthdays or adjacent birthdays." One natural extension asks for the probability of at least k people with either all the same birthday or birthdays falling within n adjacent days. At the time, this was an unsolved problem that I and others subsequently worked on.

A related problem, 26: The Hurried Duelers, finds the probability that two random points on the unit interval fall within a distance d of each other. This leads to a wealth of interesting extensions and generalizations. One generalization deals with the probability that, given N points distributed at random over the unit interval, there exists no subinterval of length d that contains at least k points. A variety of generalized cluster, coverage, and run problems grew out of studying this question. One interesting application of this generalization arose many years later. Fred was studying a variety of models and approaches for evaluating the representativeness of samples. The distribution of a measure of representativeness was derived using approaches developed in the course of solving the generalization to the Hurried Duelers problem.

Problems 47 and 48 on the Dowry Problem fall in a class of "Secretary Problems," in which Fred Mosteller gave major solutions, and which has led to a variety of other related problem areas. The reader of the book will find a rich source of inspiration.

And Marvin's mother? She should get a copy of *Fifty Challenging Problems in Probability*, turn to page 39, and tell Marvin to take the IND.

On Equality of Educational Opportunity

Frederick Mosteller and Daniel P. Moynihan, editors. *On Equality of Educational Opportunity.* New York: Random House, 1972.

by
Richard J. Light

This outstanding book tells a lot about Fred. It illustrates his work style. It represents a major contribution to a substantive field—education policy. And it shows how Fred capitalizes on an opportunity to expand his, and others', repertoire for dealing with messy but real statistical problems.

What background led to this project? In 1964, the U.S. Civil Rights Act commissioned a large national education survey. A team of researchers was assembled under the direction of James S. Coleman, then at Johns Hopkins University. Their charge was to assess the test scores, school resources, and family backgrounds of students throughout the nation. Coleman organized his team quickly and efficiently. Then in 1965 they moved into the field, testing 570,000 students and 60,000 teachers. Facilities of 4,000 schools were tallied. Data tapes were organized and analyzed, and a final report in two volumes was issued publicly in 1966.

The findings rocked the American education establishment. The survey sponsors and many educators were sure that severe inequalities in school facilities would show up. They also expected these resource inequalities to connect with differences in student performance. But Coleman's work turned up some surprises.

First, for any region of the country, school facilities available to blacks differed little from those available to whites. Where differences showed up, they were very small. For example, salaries of teachers were nearly identical for both races nationwide. Sometimes differences even favored minority students. For example, blacks in elementary schools were more likely than whites to have a full-time librarian. Blacks nationwide had more experienced teachers than whites.

Second, differences *among regions* of the country in school resources were far greater than differences among schools *within* any region. For example, the South consistently had poorer facilities than the North.

Third, achievement scores on standardized tests varied dramatically among different ethnic groups. Orientals scored well. Blacks scored below Hispanics, who in turn scored below whites.

Fourth—the most controversial finding of all. The survey found that, after controlling for student backgrounds, schools' resource differences explained less than one percent of the variation in student performance. Few expected this finding. It was not easily accepted in the education community. How could resources matter so little? Resources are controllable by policy change, and if they are not highly related to student performance, it becomes harder to think about how to improve performance.

The Coleman group's work illustrates some dilemmas that face social scientists whose work might influence policy. Despite the enormous sample size, which some observers applaud as giving the findings immediate face validity, several methodological dilemmas make it difficult to interpret findings. For example, measurement problems immediately turned up—some data on children's family background were reported by the children themselves, and thus were not entirely reliable. Another example involves the regression analyses. The goal was to relate home and school factors to children's achievement. Yet each of these, home and school, is related individually to children's achievement. In addition, these two background factors are themselves related. Deciding how to disentangle these associations is a thorny dilemma. The result was heavy criticism of Coleman's work, some of it from the social policy community and some from statistical and other scientists concerned about methodological adequacy.

Organizing a Faculty Seminar

Stimulated by this controversy about a major policy question, Pat Moynihan and sociologist Tom Pettigrew decided to sponsor a faculty seminar. They aimed to gather people from different disciplines for reanalysis of the Coleman data. When the Carnegie Corporation agreed to support their venture, and Fred began to take leadership in organizing the reanalysis, the seminar quickly became a major activity at Harvard. I think that Fred himself was a bit surprised at the breadth and depth of interest. More than 60 people participated, and about 40 were "regulars." These included such a diverse cast as statisticians Alex Mood, Howard Raiffa, and John Tukey; educators Harold Howe, Gerald Lesser, and Ralph Tyler; lawyers Abram Chayes, Frank Michelman, and Albert Sacks; sociologists David Armor, Christopher Jencks, and Nancy St. John; and many others. The participants also included several doctoral students. I especially mention Marshall S. Smith, whose major chapter in the book was part of his doctoral thesis under Fred's supervision. Smith went on to a senior position at the U.S. Office of Education under President Carter.

What is so impressive about this enterprise? At a time when many people at Harvard *talked* about the value of interdisciplinary work, Fred, Pat Moynihan, and Tom Pettigrew actually *did it.* Not only did they succeed in having people express an interest in the topic and show up for a pleasant and stimulating dinner, but they also got fifteen distinguished scholars to contribute chapters to this book.

The *process* was vintage Fred: Take an important problem, preferably one where real-world policy changes might matter. Then assemble a working group, often including people who haven't collaborated before. The third step, structuring the group effort, is hard. When dealing with established scholars, one can't just tell them what to do. Rather, the group must spend a certain amount of time developing ideas and controversies it considers important, and then shaping a work plan. For this book Fred coordinated 60 people in biweekly meetings for over a year. They produced a book that flows smoothly from start to finish.

Statistical Topics Treated in the Book

If Fred had done nothing more than organize this group, it would have been a major contribution. But, as usual, he played a broader role. Any reanalysis of a large, complex data set faces some statistical challenges. I would like to mention three this project dealt with. Subsequent projects have benefited from these efforts when facing similar dilemmas.

One topic is making causal inferences from surveys. The Coleman survey was large. Its sample included students in the sixth, ninth, and twelfth grades nationally. But there was no planned or controlled intervention. No innovations were placed in schools and assessed over time. Yet popular press accounts of the Coleman Report routinely made causal statements. The most common was that racial integration would improve black students' school achievement while not hurting whites' achievement. Mosteller and Moynihan pointed out, indeed *emphasized*, the danger of making such causal inferences. Their warnings had some effect on debate in the education community. Interestingly, they seem also to have influenced the direction of Fred's own thinking. Much of his writing over the next ten years emphasized the importance of conducting randomized field trials where a new treatment is *actually implemented* and assessed over time. In this way, policymakers could avoid the difficulties of trying to make causal inferences from surveys.

A second topic is how to choose an appropriate unit of analysis. The original Coleman Survey used individual children in all regression equations. In contrast, David Armor (one of the contributors to the volume) redid much of the work using *schools* as the unit of analysis. He found that the between-school variation of many key school resource variables, such as teacher preparation, made up about one third of the total variation due to schools and individuals.

Emphasizing the study of schools rather than individuals led Armor to another finding—that schools attended primarily by black students, with mostly black teachers, generally have lower teacher verbal performance than primarily white schools with primarily white teachers. Armor found that the *teacher* gap was nearly as wide as the achievement gap between black and white *students*. He concluded that, although it is politically uncomfortable to argue the point, black students may do less well in developing verbal achievement if taught by teachers who themselves are not strong.

This question of what policymakers can learn by using different units of analysis must also have struck a responsive chord for Fred—ten years later Judith Singer produced a solid Ph.D. thesis on this exact question.

A third topic is how to interpret findings from different multiple regression analyses that enter or remove variables from equations using different orders. Any statistics student who has taken a course in regression knows that forward stepwise regression can yield different findings from backward stepwise regression—unless all combinations of variables are analyzed, usually an impractical proposition. As part of their reanalysis Fred and his colleagues, especially Marshall Smith, tried various ways of disentangling groups of variables in regressions. One group of variables contained characteristics of an individual student. A second group contained characteristics of *other students* in a school. A third group was characteristics of teachers and school resources. The reanalysis examined the amount of variation in student achievement attributable to family background, rather than to other sources of variation such as school resources. They found that the original Coleman calculations had used several incorrectly coded variables. Yet they found, to their, Coleman's, and almost everyone else's surprise, that their corrections actually *increased* the importance of family background. So the original Coleman finding about the importance of home characteristics was strengthened considerably. In addition, a fascinating new finding emerged. The importance of home background as a predictor of school achievement was *far* greater for whites than for blacks.

The Book's Impact on Practice

When Coleman's original survey findings came out, many in the professional education community simply didn't believe them. The survey came precisely at the time education reformers hoped that spending more money would improve educational achievement for American students in general, and close the gap between blacks and whites in particular. Throughout the late 1960s a vigorous debate ensued among educators, and in government, about the validity of the findings from the massive survey. The original report had so many surprises that even some of its supporters were unsure of how firmly to believe its findings.

The Mosteller-Moynihan-led reanalyses succeeded where so many academic projects fail. They advanced the debate about home versus school resources so that, whatever a reader's initial view, there is now a far more systematic methodological treatment, all pulled together in a particularly readable volume. In summary, the book's main conclusions were:

1. Within any region of the country, school facilities for blacks and whites are roughly equal.

2. Blacks begin school with lower achievement scores than whites,

and the gap grows over time, even if both races attend schools with similar resources.

3. The most important predictor of how well students do in school is their home and family background. A somewhat less important factor for any one particular student is the group of "other" students attending the same school. School facilities are the least important of all.

Large sample surveys such as Coleman's are very useful for describing the state of events at a certain point in time. But they are poor at telling us what will happen if new policies are tried. For example, the data from 570,000 students tell us nothing concrete about what increasing racial integration will do to black students' achievement. The only serious way to estimate the effect of integration on achievement is, according to Mosteller-Moynihan, to try it. And preferably with a randomized, controlled, field study. The authors are fully aware of the many possible pitfalls facing any large-scale randomized field study. But in the end, they argue, "the random assignment of the experimental treatment triumphs over all the little excuses for not attributing the effect to the difference in treatment." (p. 372).

On Equality of Educational Opportunity has played a very big role in enhancing our understanding of how school resources are distributed in America. It is one of the most widely quoted books on education policy over the past dozen years. And here at Harvard, at least, it gets the ultimate compliment. I have recently attended several planning meetings where a group of faculty colleagues wanted to organize a work seminar around a public policy question. At each of these meetings someone remarked, "If only we could get something going as exciting and effective as the Mosteller-Moynihan project."

Federal Statistics

Federal Statistics: Report of the President's Commission, Volumes I and II. W. Allen Wallis, Chairman, and Frederick Mosteller, Vice-Chairman, of President's Commission. Washington, D.C.: U.S. Government Printing Office, 1971.

by
Yvonne M. Bishop

How the Report Came About

One of Fred Mosteller's great interests is statistical policy. Thus, it is not surprising that he was vice-chairman of the President's Commission on Federal Statistics. The commission was chaired by Allen Wallis and had twelve other members. In the short space of a year, they obtained input from 39 consultants and contributors and produced a remarkable two-volume report.

In his August 1970 letter to Allen Wallis, President Nixon asked three basic questions and elaborated his concerns in each area:

> (1) What are the present and future requirements for quantitative information about our society? (2) How can we minimize the burden on respondents and insure that personal privacy and data received in confidence are protected? (3) How can we organize Federal activities for the most effective compilation and utilization of statistics?

The commission recognized that in the time available they would be unable to diagnose shortcomings or prescribe improvements for more than a tiny fraction of the federal statistical system. As explained in the letter forwarding the report, they sought instead

> to strengthen the system's capacity for self-improvement and self-renewal: its ability to recognize its own faults and to remedy them by meeting new needs, halting useless activities, adopting new methods, improving its personnel and organization, and enhancing the effectiveness with which statistics are applied to operations and policy-formulation in the federal government.

Contents of the Report

The first volume of the report is organized to stress the theme of strengthening what exists. The initial chapter, "Issues and Opportunities," describes selected federal activities via brief anecdotes illustrative of the diversity of what exists. Then a description of some major achievements of the federal statistical system leads to the main message of the report:

> While major changes in the federal statistical system are not called for, rich opportunities abound for extending past progress.
>
> A primary task confronts those who use and those who produce federal statistics: to improve the quality of these statistics in the decade ahead. This is the single most urgent need facing the federal statistical system.

The suggested areas of potential improvement include measurements of error, winnowing out the trivial in order to conserve resources for the important, application of sampling instead of complete enumeration, using statistical methods to evaluate government agencies, and finally the need to coordinate statistical activities.

The next two chapters describe respectively the producers and users of statistics. The emphasis is on the decision-making process, and examples are given of the results of the process:

> although producers of statistics normally do not set the time limitations or the definitions to be used, they are often blamed if the data produced under someone else's specifications result in bad decisions.

Specific recommendations are mostly given in appendices. "Some New Directions," an appendix to chapter two, has ideas related to the measurement of prices, methods of classification, the need to develop a set of "social accounts," the need to improve crime and law enforcement statistics, and finally the importance of measuring change. Chapter 3 has one appendix on program evaluation and another on the federal role in providing small area data.

Repeated themes in these chapters and the following chapter on "Organization and Coordination" are the role of the Statistical Policy Division (SPD) in the Office of Management and Budget (OMB), the effect of the budget process on data coverage, and the difficulties of identifying both gaps and overkill. The achievements of SPD are described and further activities suggested:

> Modern statistical techniques and procedures can be used much more fully than they are now being used to inform policy-makers of the nature of the problems they face and the conse-

> quences of the decisions they make. Basic to the use of modern practice is the quality of personnel in the system, ... In addition to the present activity of publishing its journal, the *Statistical Reporter*, SPD could increase communication and the feeling of community among government statisticians by organizing a central program of seminars and on-the-job training.

A suggestion that the statistical agencies should be subjected to statistical audits is more fully developed in the appendix to this chapter that gives the purpose, scope, and specific objectives of such audits and even provides a checklist for conducting them.

The summary chapter on "Findings and Recommendations" explicitly recommends that a National Academy of Sciences-National Research Council (NAS-NRC) committee be established to review federal activities and report to the Director of OMB. The subsequent recommendations are more specific and deal with the need for directives, linkages, provision of access to unit data, and the creation of an all-agency catalogue. An interesting addition is the provision of an Assessment Checklist to be used when a new program is being considered. This is particularly pertinent today, when the Paperwork Reduction Act forces many trade-offs and creates a new emphasis on cost-benefit analysis of needs.

The second part of Volume 1 of the report deals with the issues of privacy and confidentiality, complex issues that hamper the exchange of information between agencies. Debate and discussion on these unresolved issues still continue today.

Of the twelve signed chapters that make up Volume 2, only one was authored by a member of the commission—Chapter 6, "Using Controlled Field Trials to Improve Public Policy," by Richard J. Light, Frederick Mosteller, and Herbert S. Winokur, Jr. This seems a typical example of Fred's indefatigability. Not only was he exceedingly active in producing the main report, but he also found time to work in more detail on an aspect that seemed particularly important to him.

Were the Recommendations Effective?

Reviewing these recommendations after 14 years, it is natural to ponder their impact. Of the more general recommendations, the advice on enlarging the functions of SPD has not been followed as envisaged. This group's disjointed history includes a transfer from OMB to Commerce in 1977 and a return to OMB in 1981 as a branch within the Office of Information and Regulatory Affairs (OIRA), as described by James Bonnen (1983). With reduced resources the ongoing *Statistical Reporter* was discontinued, and suggestions for additional functions, such as Federal statistical training, were not adopted. The need for such centralized functions has nevertheless been recognized by other organizations. The Washington Chapter of the American Statistical Association (ASA) has assisted interagency groups of statisticians to organize training seminars. *The American Statistician* has been generous in its coverage of matters concerning the statistical agencies, and the ASA has published reports that would naturally have been part of OMB's guideline series. There has also been an increase in the activities of private organizations that disseminate information about federal statistics, such as the Association of Public Data Users (APDU), the Council of Professional Associations on Federal Statistics (COPAFS), and the Consortium of Social Science Associations (COSSA).

The role of OMB in guiding federal statistics has been greatly influenced by the Paperwork Reduction Act of 1980, which aimed to reduce the paperwork burden on the public by 25 percent between 1980 and 1983. Thus, the recommendation for winnowing out less important data collections has certainly been followed (OMB 1985). The 1980 Act also gave OMB the responsibility for statistical policy and coordination functions, specifically long-range planning, coordination, development of policies, principles, standards, and guidelines, and evaluation of statistical program performance. The small staff has tackled these responsibilities mainly by setting up a variety of committees. Some committees give advice on a particular survey, such as the Survey of Income and Program Participation Advisory Committee, and the Federal Agency Council for the 1990 Census. Others deal with cross-agency issues such as the revision of the Standard Industrial Classification scheduled for 1987. One long-standing interagency committee, the Federal Committee on Statistical Methodology, was established in 1975 and still functions effectively; it sets up groups to study and report on methodological issues, and recently these have been followed by workshops designed to spread the findings.

Thus progress has been made along the lines that the commission suggested, although not in the manner that was envisaged. The commission, when comparing what might be done with what was being done, were aware of the difficulties and constraints in what they proposed:

> The observation that SPD has been trying to do an impossible task with an inadequate tool is not intended as a criticism of SPD. The tool itself has had a profound influence on the perception of what can be done.

The suggestion that agencies be audited has borne some fruit. It seems likely that this recommendation influenced Congress in 1976 when the Office of Energy Information and Analysis was set up within the Federal Energy Administration. (This office was the predecessor of the current Energy Information Administration within the Department of Energy.) The enabling legislation contained the requirement that a Professional Audit Review Team (PART) should audit the statistical activities of the agency annually. The PART report is submitted to Congress and is publicly available. This mechanism has helped the agency by providing Congress with an impartial report on the quality of the data. Congress has consequently provided funds for quality improvement.

As recommended by the commission, in 1972 the Committee on National Statistics was established under aegis of NAS-NRC. The first chairman of the committee was William H. Kruskal, a member of the commission, and Fred Mosteller was one of the original committee members. Subsequent chairmen have been Conrad Taeuber, Lincoln Moses, Stephen E. Fienberg, and Burton H. Singer; the work of the committee has been summarized in reports in *Science* (Kruskal, 1973), *The American Statistician* (Martin, 1974, 1976; Martin et al. 1982), and elsewhere. The appreciation of the federal agencies is shown in a concrete way: they provide funds for the committee through a consortium arrangement and frequently fund special studies that are carried out by panels set up by the committee. Some studies have taken the form of a review of a statistical agency's program, such as the "Review of the Statistical Program of the Bureau of Mines" (1982), or have dealt with particular aspects of an agency's program. Other studies have examined issues cutting across agencies, such as "Privacy and Confidentiality as Factors in Survey Response" (1979) and "The Comparability and Accuracy of Industry Codes in Different Data Systems" (1984).[5] The committee has also sponsored conferences, including one on Indicators of Equity in Education in 1979 and another on Immigration Statistics in 1980. A listing of the committee's publications shows that they have followed up on many of the concerns that the commission expressed. Their independent reports provide great assistance to the federal agencies.

In this review, I have mentioned some outcomes that can be ascribed to the commission's impact. In other instances, I have not considered what subsequently happened; to do so would probably result in another two-volume report! It seems that the commission members were primarily concerned that the results of their effort should not be just another report that languished on the shelf. Instead they focused on the need to establish ongoing review mechanisms. The effects of this recommendation continue today. A report by a commission seldom identifies the contributions of the particular individuals. This report, however, has certain features that Fred has sought to incorporate in any endeavor with which he has been involved. Throughout the report care has been taken to focus on the main issues, and relegate to appendices the details of both the findings and the recommendations. Thus, the report is easy to read. The main points are illustrated with varied cogent examples. There are many checklists—I remember Fred's remark, when confronted with a checklist is an entirely dif-

[5]Publications cited are available from the National Academy Press, National Academy of Sciences, 2101 Constitution Avenue, N.W., Washington, D.C. 20418.

ferent context, "Your readers will love it." Finally, no review of the report would be complete without mention of the light-hearted cartoons throughout; I was particularly delighted in the chapter on confidentiality by the sight of a camera trained on a person in heavy veils.

I remember when Fred returned from presenting the report to President Nixon. He was elated with the President's understanding reception of it and remarked, "It was like talking with a colleague." I hope he is as pleased today with the impact it has had.

References

Bonnen, James T. (1983). Federal statistical coordination today: A disaster or a disgrace? *The American Statistician*, **37**, 179–202 (with discussion).

Kruskal, William (1973). The Committee on National Statistics. *Science*, **180**, 1246–1258.

Martin, Margaret E. (1974). The work of the Committee on National Statistics. *The American Statistician*, **28**, 104–107.

Martin, Margaret E. (1976). Report on activities: Committee on National Statistics. *The American Statistician*, **30**, 21–23.

Martin, Margaret E., Goldfield, Edwin D., and Straf, Miron L. (1982). The Committee on National Statistics: 10 years later. *The American Statistician*, **36**, 103–108.

Office of Management and Budget (1985). Managing Federal Information Resources. Fourth Annual Report under the Paperwork Reduction Act of 1980. September 1985.

National Assessment of Educational Progress

Reports for which Frederick Mosteller was coeditor:

Report 2. *Citizenship: National Results.* November 1970. Washington, D.C.: U.S. Government Printing Office.

Report 3. *1969–1970 Writing: National Results.* November 1970. Washington, D.C.: U.S. Government Printing Office.

Report 2–1. *Citizenship: National Results.* November 1970. Education Commission of the States, Denver, Colorado and Ann Arbor, Michigan.

Report 4. *1969–1970 Science: Group Results for Sex, Region, and Size of Community.* April 1971. Washington, D.C.: U.S. Government Printing Office.

Report 5. *1969–1970 Writing: Group Results for Sex, Region, and Size of Community.* April 1971. Education Commission of the States, Denver, Colorado and Ann Arbor, Michigan.

Report 7. *1969–1970 Science: Group and Balanced Group Results for Color, Parental Education, Size and Type of Community and Balanced Group Results for Region of the Country, Sex.* December 1971. Education Commission of the States, Denver, Colorado.

Report 8. *Writing: National Results—Writing Mechanics.* February 1972. Washington, D.C.: U.S. Government Printing Office.

Report 9. *Citizenship: 1969–1970 Assessment: Group Results for Parental Education, Color, Size and Type of Community.* May 1972. Education Commission of the States, Denver, Colorado.

Report 02–GIY. *Reading and Literature: General Information Yearbook.* May 1972. Education Commission of the States, Denver, Colorado.

by
Lyle V. Jones

From 1970 to 1976, Frederick Mosteller served the National Assessment of Educational Progress (NAEP) as a member of its Analysis Advisory Committee (ANAC). From 1973 to 1976, he chaired that committee and was a member of the NAEP Policy Committee. Mosteller's bibliography cites nine NAEP reports published between 1970 and 1972 for which he was a coeditor. While his contributions to those reports and to several other NAEP reports were substantial, his contributions to the philosophy and the development of the NAEP project were at least as great. I shall try here to document both forms of influence.

NAEP, as a concept, began taking shape in the early 1960s. With funding from private foundations and active leadership from Ralph W. Tyler, the Exploratory Committee on Assessing Progress in Education (ECAPE) began work in 1964–1965. Representing ECAPE, Ralph Tyler in 1965 asked John W. Tukey to chair ANAC (originally TAC, the Technical Advisory Committee), which would assume primary responsibility for the design of a national assessment program and for the analysis and reporting of its findings. Charter members of TAC/ANAC in addition to Tyler and Tukey were Robert P. Abelson, Lee J. Cronbach, and Lyle V. Jones. This group remained intact until 1969—the year of the first data collection for NAEP—by which time it had designed a long-range plan for the periodic assessment of achievement levels of the nation's youth, especially emphasizing change in achievement from one assessment year to another. NAEP received funding from the U.S. Office of (Department of) Education, and the project was managed from 1969 until 1982–1983 by the Education Commission of the States.

In 1970, Ralph Tyler and Lee Cronbach retired from ANAC; new members were William E. Coffman, John P. Gilbert, and Frederick Mosteller. By 1976, the membership was Mosteller (Chairman), David R. Brillinger, Janet Dixon Elashoff, Gene V. Glass, Lincoln E. Moses, John Gilbert, Lyle Jones, and John Tukey. (James A. Davis was an ANAC member in 1973–1974. By 1982, ANAC membership consisted of Janet Elashoff as chairman, R. Darrell Bock, Lloyd Bond, Jones, and Tukey, at which time the Educational Testing Service replaced the Education Commission of the States as NAEP contractor, and ANAC was retired.)

Prior to 1969–1970, ANAC had developed detailed prescriptions for NAEP target populations, achievement areas to be assessed, exercise development criteria, exercise administration procedures, stratified sampling designs, and data-analysis and reporting strategies. The plan included a number of novel features peculiar to the ambitious aims of a national educational assessment. Therefore, ANAC agreed to assume primary responsibility for preparing a round of reports for one achievement area that might serve as a model for the data analysis and reporting of results in other areas. The area selected was science; science was assessed in the first NAEP survey of 1969–1970 and was the first area for which change results would become available (in 1973–1974 and again in 1976–1977). ANAC, then, contributed directly, with NAEP staff assistance, to NAEP Reports 1 (1970), 4 (1971), 7 (1971), 04–S–00 (1975), and 08–S–21 (1979). It also was involved to a lesser extent in other NAEP reports.

Reports 4 and 7 required the charting of especially treacherous waters. The texts of these reports were prepared in an extended series of telephone conference calls involving Mosteller, Tukey, Abelson, Coffman, Gilbert, and Jones, typically beginning around 8:00 p.m. and ending well after midnight.

For this group of participants, the device worked well, not only for detailed editing, but also for the insertion of new text and the design of tables and graphs, probably because the group was thoroughly acquainted from earlier meetings and because each member freely expressed his views while also respecting the judgments enunciated by the others.

Fred Mosteller provided key contributions to the solution of sticky issues that arose during this preparation of the NAEP reports. Characteristically, during a pause following a prolonged argument between other ANAC members about an issue, he would offer a soft-spoken, articulate suggestion that not only resolved the conflict but also both simplified and enriched the text. His fine hand is evident in many passages from these reports, two of which are reproduced here. From Report 4 (1971):

> There is a kind of interpretation that should never be made on the basis of the sort of figures given in this report. The fact that figures reflect Southeast performance or Big City performance does not mean that the performances thus reflected have arisen precisely from living in the Southeast or in a Big City, or from the attitudes, techniques, facilities and staffs of the school system involved. In particular, just what happens in a region involves other things than that region's schools. Larger fractions of the children in some regions belong to a particular size-of-community group. Thus effects due only to size of community can appear to be regional differences. Larger or smaller fractions of the parents in some regions have particular amounts of education. Thus effects due only to parental education can appear to be regional differences. And so on. Migration from one region or size of community to another can further complicate the picture. There are such difficulties, some of which we know how to adjust for, and some of which we do not.

From Report 04–S–00 (1975):

> A number of factors might account for declines in science achievement between 1969–1970 and 1972–1973. The decline in the number of young Americans demonstrating knowledge in science may simply reflect the fact that fewer students are electing to take science courses. It is possible that achievement levels may be rising among those students enrolled in science curricula; the National Assessment of Educational Progress (NAEP) assessed *all* students at these age levels, not just those interested in science. Teachers may well be doing a better job than ever with a select group of students who will form the nucleus of the next generation's scientific community.

Perhaps the results of the science assessment are satisfactory and as expected. Quite possibly less attention is being paid to science in the schools these days; the declines between 1969 and 1973 may only suggest that unusual emphasis was given to science education in the wake of Sputnik, and the new results reflect a return to more normal achievement levels. On the other hand, our society is becoming more technological and complex, and this may be a poor time for such a decline. Average citizens must have some basic knowledge about science to remain informed about critical social and environmental issues upon which they are expected to make rational decisions. Can the average citizen know too much science, given the complexity of the world that confronts us? Should the general public be encouraged to appreciate science for its own virtues? Are we training enough good scientists to meet the growing needs of our technological society? Should scientists and science educators attempt to make science more relevant to the social and political concerns of both majority and minority Americans? The public at large should join the scientific and educational communities in discussing these items.

From its inception, ANAC regularly discussed means by which NAEP results would be disseminated. With some naivete, the Committee envisioned the possibility that newspaper columnists would daily interpret children's achievement levels on individual NAEP exercises, applauding or decrying a result depending on whether it surpassed or failed to meet an expected level of performance. The Committee anticipated a keen interest in NAEP findings on the part of parents, educators, and the general public, and was disappointed when evidence of intense interest failed to materialize. Mosteller, especially, challenged the committee to discover ways in which NAEP results could be published to "make a difference" in education. With active collaboration and support from the late John Gilbert, he persistently encouraged ANAC, the NAEP Policy Committee, and the NAEP staff to remain alert to the essential need for policy implications of NAEP findings in order to justify continuation of the project. These interests, the breadth of Mosteller's vision, and his sense of societal imperatives were documented in 1972 by a position paper prepared by Mosteller and Gilbert, "National Assessment in the National Interest: Perspectives on the First Round of Testing," which contains the following selected passages.

Informing the General Public

Because of the wide concern with educational achievement, NAEP must inform the public about the level of performance of the nation on the great variety of questions we ask. It is important that this information be presented in a newsworthy way so that the mass media can pass along the information. This effort at communication has been going on in the NAEP newsletter, in the releases, and in the press conferences, but now the opportunity arises to present the findings of the whole first round. A more elaborate news story can now be made of the several reports written so far, cast in a way that newswriters will take advantage of for their own stories.

Contributing to the Disciplines

NAEP can commission a number of specialists in disciplinary fields to write papers on the value and uses of the work done by NAEP for their particular discipline. What new or old facts has NAEP brought? What information do the data in the reports bring that subject matter specialists should be considering? Is there further research into the data that the subject matter people should do? Or does the information suggest research outside NAEP that others should do? One reason we need subject matter specialists is that they can put the material in a setting and give it some thrust. And so all told the questions are: What did we know before? What is new here? and, What research is needed to find out more?

Can Basic Research Workers Help?

In addition to the general public, the disciplinary teachers, and the disciplinary research specialists, we might also attract the interest of the research specialists in the parent disciplines to supplement the work of the applied disciplines. One job of considerable importance to which NAEP has made an almost unique contribution is to measure the size of differences and not just their direction. Are the differences between groups that we now observe comparable to those in the 1966 Coleman report? Or are they declining?

National Policy

In the long run the Office of Education and Congress must take an interest in the efforts of NAEP. On the one hand, the kinds of papers suggested above would form a foundation of information for the Office of Education and for testimony before the appropriate congressional committees. On the other hand, there are many questions that Congress could well ask that are not likely to be answered by the information already described. Congress needs in the first instance information on the value of NAEP to the field of education as a whole and secondly on its value for public policy, the funding of education, the directions new programs should take, the opportunities to strengthen education in minority groups, and the research required if our educational monies are to be well-spent. Some task force is required to explain both the contributions of NAEP to these problems and the limitations of its results. The explanation should also suggest further questions whose answers are needed if the nation is to progress. Therefore, we need to find a political statesman who appreciates the role of Congress as well as the value of concrete findings to write a paper of value to the Congress. An example of some interest to Congress might be the clear documentation of inequalities in educational opportunity when we compare affluent suburbs with inner city schools.

The States and the School Districts

Naturally a parallel document needs to be prepared to explain to the States and to the school districts the contributions of NAEP, both those already achieved and possibilities from follow-up investigations that local and state groups may wish to make. For example, examinations could be given that include some of the released questions to see how well groups in the particular districts perform compared to the performance in the region and to the size and type of city.

At the moment we have no direct way of relating measures of achievement of students with measures related to the school, for example, its disciplinary style.

The Evolution of NAEP

We would anticipate that the staff could reach into its experience with government, the press, the States, and the school districts as well as the public to explain the concerns these groups have about NAEP, the problems they hope NAEP can help solve, and NAEP's role in the actual event. What does it seem that the future will require? What steps will be needed?

Inevitably NAEP itself would be a consumer as well as a producer in this program. It will want to examine the deeper implications of what it has accomplished. The many discussions will inevitably suggest to the staff and Policy Committee new sorts of studies and services for NAEP to consider. And the discussions should create an atmosphere conducive to deciding again where NAEP responsibilities leave off and those of others begin. We view this exercise as partly a device to help NAEP think about its past and future contributions, its opportunities

> and its limitations. And, of course, we view it as a device for consolidating and disseminating the information already gained.

These suggestions were endorsed by ANAC and by the NAEP Policy Committee; although many were implemented and contributed directly to a greater visibility and wider utility of National Assessment findings, ANAC members continued to be perplexed by the difficulty of devising more effective ways for disseminating reports.

Mosteller's contributions to NAEP developed as a consequence of a combination of his interests, skills, and talents: a keen interest in effective education, a mature facility for research design and data analysis, sensitivity to the importance of policy-relevant research and insight into classes of findings that are policy-relevant, exceptional writing skills, interpersonal warmth, and genuine good humor. Neither his influence on NAEP nor on broader issues in education ended when he retired from active participation in the NAEP project. His successor as ANAC chairman, John Gilbert, maintained the emphasis on the policy relevance of NAEP, while Mosteller's efforts to improve educational effectiveness have continued in several other projects, whose sponsors include AAAS, ASA, the National Science Board, and the Consortium for Mathematics and Its Applications.

Statistics: A Guide to the Unknown

Judith M. Tanur and members: Frederick Mosteller, Chairman, William H. Kruskal, Richard F. Link, Richard S. Pieters, and Gerald R. Rising of the Joint Committee on the Curriculum in Statistics and Probability of the American Statistical Association and the National Council of Teachers of Mathematics, editors.
Statistics: A Guide to the Unknown. San Francisco: Holden-Day, 1972. Second edition: 1978. With Erich L. Lehmann, Special Editor. Re-issued: Monterey, CA: Wadsworth & Brooks/Cole, 1985.
Statistics: A Guide to Business and Economics. San Francisco: Holden-Day, 1976. With E.L. Lehmann, Special Editor.
Statistics: A Guide to the Biological and Health Sciences. San Francisco: Holden-Day, 1977. With E.L. Lehmann, Special Editor.
Statistics: A Guide to Political and Social Issues. San Francisco: Holden-Day, 1977. With E.L. Lehmann, Special Editor.

by
Gudmund R. Iversen

"Actually, quite interesting," is an evaluation my students commonly give *Statistics: A Guide to the Unknown* at the end of their introductory statistics course. The book has been used for several years as a supplement to the regular textbook, and the students often volunteer their opinions about it.

At the beginning of the semester I tell the students that the final examination will contain a question from the book. They may bring any notes they want, but not the book itself, to the examination. I advertise the book as an excellent and well-written collection of examples of uses of statistics. Also, nobody will benefit from my going through these examples in class since the book is so good, and so the students are on their own reading the book. All I want is that the students read the book. Ideally they would do it without the threat of an examination question at the end, but I have not found a way to achieve this ideal yet. Perhaps no book on statistics is that interesting to nonstatisticians.

Frederick Mosteller chaired the joint committee of the American Statistical Association and the National Council of Teachers of Mathematics that produced the book, and it is a considerable delight to report that students find the book "interesting." When I tried to use that word in an early manuscript, Mosteller told me in no uncertain terms to leave it out. "You just write," he said, "and I will decide, as a reader, whether it is interesting." It seems very fitting that this book has passed the "Mosteller test" in the eyes of the students. On the other hand, the students' use of the word "quite" may indicate that they have not read and fully comprehended all the material.

The book grew out of the curriculum studies and reforms that took place in the 1950s and 1960s in secondary schools and colleges. Both mathematicians and statisticians wanted to broaden the appeal and understanding of their disciplines, not only among students but also among the general public. *Statistics: A Guide to the Unknown* is a direct result of these concerns. The book was seen as a way of explaining, through well-chosen examples, why statistics needed to be taught to more people and why contributions made by statisticians are important to society.

Several things come together to make *Statistics: A Guide to the Unknown* and the three spinoff volumes such excellent books. One is the talents of the more than fifty authors who wrote the 46 chapters. The list of authors reads like a Who's Who of statistics. It must have taken considerable persuasion to get so many talented people to write for a volume which, in the beginning, did not have a publisher and whose future sales were uncertain.

Another reason for the excellence is that the topics in the various chapters span such a wide range of applied fields. The four parts of the main book cover the biological, political, social, and physical world around us. This framework later made it easy to publish subsets of the chapters in three smaller, separate volumes. Without such a wide range of applications, the books would have had much more limited appeal.

A random sample of chapters discuss the Salk polio vaccine trial, deathdays and birthdays, parking tickets and missing women jurors, the relationship between registration and voting, estimating demand for a new product, the consumer price index, racial integration, and the role of statistics in studying the sun and the stars. Such a great variety of topics allows almost any reader to find something of particular interest, and this encounter leads the reader to other chapters.

But the combination of talented people and good topics does not automatically produce a readable book on the many uses of statistics for the person who has only a limited background in quantitative thinking. It is often easier to write for a peer than for someone who knows much less than we do. It is to the great credit of Mosteller and his coeditors that they succeeded so well in making this book so accessible to nonstatisticians.

What makes the chapters so uniformly readable seems to be that each author explains the simple statistical features of the study and does not necessarily try to explain more sophisticated features. It takes strong editorial persuasion and power to get writing like that from such a large group of authors. Most often we feel no need to explain what we think is obvious; we would rather explain the new and path-breaking. But the uninitiated reader often finds such simple and obvious matters new and challenging, and the editors understood this point from the very beginning.

It is one thing to ask authors for this kind of writing and another thing entirely to show that it can be done. With this in mind, Lincoln Moses and Fred Mosteller prepared their chapter on safety of anesthetics early on, and this chapter was sent to each author as an example of what the editors wanted in level and style. But, knowing how hard it is to write, I believe the editors still had to put in a great deal of work before the chapters became the excellent book we have today.

In spite of the many topics, the main book is very well organized and has unusually extensive and helpful tables of contents. First, the basic table of contents lays out the four parts: biological, political, social, and physical. Within each part subheadings group the chapters, and after the title and author each chapter has a one-sentence summary. A second list arranges the titles of the chapters according to their main sources of data: samples, available data, surveys, experiments, and quasi-experiments. Finally, a third list of contents uses types of statistical tools as a basis for organizing the chapters. The eleven groups of tools range from inference to decision making. Together, these three schemes for listing the chapters make it possible to approach the book and see patterns in a variety of ways. The book also includes an extensive index.

In addition to the chapter on the safety of anesthetics with Lincoln Moses, Mosteller himself participated in the writing of three other chapters: "How frequently do innovations succeed in surgery and anesthesia?" with John Gilbert and Bucknam McPeek, "How well do social innovations work?" with John Gilbert and Richard Light, and "Deciding authorship" with David Wallace. The first two chapters arise from his interest and involvement in the large study of the effect of halothane as an anesthetic. The third chapter comes from his long interest in the use of statistics in the social sciences and the setting of public policy. The fourth chapter is an outgrowth of the study he and Wallace made of the *Federalist* papers.

Together, these chapters give us a glimpse of Mosteller, not as an editor, but as a statistician with a broad range of interests in the use of statistics. Without such applications, statistics would remain a minor branch of mathematics. But, thanks to the efforts of people like Frederick Mosteller, statistics has become what it is today, a field of inquiry that modern society cannot be without.

Sturdy Statistics

Frederick Mosteller and Robert E.K. Rourke. *Sturdy Statistics: Nonparametrics and Order Statistics.* Reading, MA: Addison-Wesley, 1973.

by
Ralph B. D'Agostino

Sturdy Statistics is a gem. It blends elementary and sophisticated concepts in a careful manner that allows the reader to absorb substantial amounts of both levels with only minimal awareness of the difficulty of some of the material. For the most part the reader should find the book fun to read.

The non-standard title, *Sturdy Statistics*, only hints at the book's content. The subtitle, *Nonparametrics and Order Statistics*, is more explicit and reveals better, but still not completely, the content of the book. The preface solves the mystery. "Sturdy" conveys the same ideas as the technical terms "robust" and "resistant," but it avoids giving some potential readers the impression that the book contains a study of comparative robustness of various statistics. The term "sturdy statistics" refers to "methods that retain their useful properties in the face of changing distributions and violations of standard assumptions." The book emphasizes "the statistics themselves and their applications, as is appropriate in an early course."

The intended audience is those with two years of high school algebra, along with some exposure to statistics. Seven appendices help to recall or reinforce the needed statistical prerequisites. These review the contents of a standard elementary statistics course covering discrete and continuous random variables, the central limit theorem, covariance, correlation, sums and differences of random variables, sampling theory, and statistical inference notions from hypothesis testing and estimation through inferences on the difference of two population means. The appendices are formal with an abundance of definitions, statements of theorems, subscripts and occasionally an integral sign implying some knowledge of calculus. Numerical examples, spread liberally throughout, illustrate and reinforce the formal concepts. It is doubtful that the appendices could be used solely to learn and understand elementary statistics. They do, however, serve well to refresh and reinforce the reader. They offer a good review. My personal feeling is that the appendices are harder than the main parts of the book. A previous course, more elementary than is implied by them, is sufficient.

The book proper has sixteen short chapters, ranging from five or six pages up to a maximum of twenty-four. The table of contents shows the breadth of coverage:

1. Normal and Non-normal Data

2. Applications of the Binomial and Multinomial Distributions

3. Ranking Methods for Two Independent Samples

4. Normal Approximation for the Mann-Whitney Test

5. Wilcoxon's Signed Rank Test

6. Rank Correlation

7. Rare Events: The Poisson Distribution

8. The Exact Chi-Square Distribution

9. Applying Chi-Square to Counted Data: Basic Ideas

10. Applying Chi-Square: $1 \times k$ Tables

11. Applying Chi-Square: Two-Way Tables

12. The Kruskal-Wallis Statistic

13. The Problem of m Rankings: The Friedman Index

14. Order Statistics: Distributions of Probabilities; Confidence Limits, Tolerance Limits

15. Order Statistics: Point Estimation

16. Designing Your Own Statistical Test or Measure

The first thirteen chapters cover the usual topics of nonparametric statistics. The level is elementary but clearly more advanced than S. Siegel's *Nonparametric Statistics for the Behavioral Sciences* and G. Noether's *Introduction to Statistics.* Examples introduce concepts. Exercises reinforce them. The examples are both simple and complex, some more complex than is usual in texts at this level. These show the usefulness and applicability of these procedures.

In addition to the usual presentation of nonparametric tests and extensive tables accompanying them, these chapters contain three unique topics usually not in these types of books. They are permutation tests, power of statistical tests, and maximum likelihood estimation. All are well presented and reinforced throughout the chapters. The sections on permutation tests introduce the reader to a concept of broad generalizability. One of Professor Mosteller's talents, clear to anyone who has the occasion to work with him, is his ability to explain profound concepts with simple illustrations. The discussions of permutation tests provide a good example. The new reader is bound to benefit greatly from them.

The sections on the power of various nonparametric tests supply a long needed feature. Often these tests are presented in a grab bag fashion with theory at best only at an intuitive level. Such an approach supplies no sense of alternative hypotheses and sample size considerations. Mosteller and Rourke avoid this and present discussions of power that help the reader to understand these techniques as valid useful tools for research. Such an understanding is hard to achieve when the techniques appear only as fillers in the last chapter of an elementary text.

The introduction of maximum likelihood estimation at this level is more successful than I would have believed possible. A mixture of graphical analysis, algebra, and calculus illustrates this important notion. The reader is well equipped to encounter this concept in more advanced settings.

These chapters (Chapters 1 to 13) represent a mixture of traditional topics with a liberal does of unique features, all pedagogically well done. The next two chapters depart from most elementary treatments and deal with order statistics used for confidence limits, tolerance limits, and point estimation. These chapters are exciting. The simple presentation of confidence intervals for quantiles is so enchanting one wonders why they are not used more often. The simple development of tolerance intervals makes the reader long for more. More material on confidence intervals and tolerance limits would have been welcome.

Chapter 15 on point estimation from order statistics is my favorite. The chapter states Mosteller's 1946 result on the asymptotic distribution of the order statistic U_i and uses it to illustrate how well the sample median performs as an estimate of the population median. The concept of efficiency is introduced for comparing point estimates. Trimmed means and the use of the sample range to estimate the population standard deviation are also introduced and well illustrated. All of these topics are given too briefly. One wishes for more.

Professor Mosteller's interest and direction on these topics influenced much of my own early research. Estimates of quantiles based on order statistics were my first interest. This I worked on directly under him. From this came my investigation of estimators based on order statistics for parameters of the Weibull, Rayleigh, logistic, and normal distributions. Simple robust and adaptive estimates also based on order statistics took shape at the same time. Finally my work on tests of normality has its roots in Mosteller's influence. The tests of interest to me compare estimates of the population standard deviation based on order statistics to the sample standard deviation. Chapter 15 of *Sturdy Statistics* describes the bases for these.

Chapter 16 invites the reader to try using the concepts of the book to develop a statistical test or measure. A test for trends provides an illustration.

In total, *Sturdy Statistics* is an excellent small book. The demand for a second elementary statistics course in sturdy statistics was and is probably not great. The trend to data analysis, design of experiments, and multiple regression courses is well established. Still, this is not a book to miss. I was at Professor Mosteller's house shortly after *Sturdy Statistics* was published. It fully justifies the pleasure with which he displayed it that night.

Statistics by Example

Frederick Mosteller, Chairman, William H. Kruskal, Richard S. Pieters and Gerald R. Rising, The Joint Committee on the Curriculum in Statistics and Probability of the American Statistical Association and the National Council of Teachers of Mathematics (editors), a series of 4 books:

Statistics by Example: Exploring Data.
Statistics by Example: Weighing Chances.
Statistics by Example: Detecting Patterns.
Statistics by Example: Finding Models.

Menlo Park, CA: Addison-Wesley, 1973.

by
Janet D. Elashoff

At a Harvard reunion some years ago, Miles Davis offered a salute to Fred Mosteller: "Not only has Mosteller made many important contributions to statistics, but he has made sure that everyone around him has made many important contributions to statistics." The four volumes of *Statistics by Example* (SBE) provide yet another example of Fred's skill at doing just that.

Statistics by Example contains a collection of data sets, each chosen to illustrate a statistical concept or technique. The Joint Committee on the Curriculum in Statistics and Probability of the American Statistical Association and the National Council of Teachers of Mathematics, headed by Fred Mosteller, wrote to many statisticians and asked each to contribute two examples. The completed volumes show that Mosteller and the committee, aided by a task force of teachers, did much more than assemble and edit submitted examples. They discarded examples that didn't fit, constructed new examples to fill gaps, and did considerable shaping to make each example build on others and fit into its niche. Several examples appear, treated with growing sophistication, in successive volumes.

The examples discussed in the four volumes of SBE are chosen and crafted to be interesting and accessible to the general reader. Many also have special appeal for particular audiences—biologists, psychologists, sports fans, and so on. The questions addressed in the examples are typical of frequently faced problems in many areas. Several examples capture nicely the mystery and interest of numbers. Intriguing titles such as "Prediction of election results from early returns" and "Calibration of an automobile speedometer" make the reader say "Ah, I always wanted to know how to do that." After years of looking for examples for my own teaching, I know how difficult it is to find real examples that can be briefly explained at an elementary level.

Medical researchers often ask, "What book should I read to start learning about statistics? I have no math or statistics background." I seldom know what to recommend because "elementary" statistics books often lack scientific sophistication and leave readers thinking that there is no more to the practice of statistics than plugging numbers into boring formulas. More advanced books, on the other hand, typically require considerable mathematical and/or statistical knowledge. SBE helps fill this gap between oversimplified statistical treatments and books inaccessible to scientists in other disciplines. The SBE discussions encourage the reader to think rather than try to guess which formula to apply, and they are sophisticated without requiring much "math."

A few highlights from SBE illustrate both flavor and content. The first volume, *Exploring Data*, focuses on descriptive statistics, emphasizing informative ways of displaying data to facilitate "looking at it." Yvonne Bishop's "Examples of Graphical Methods" contains scatterplots that illustrate nicely the varied effects grouping variables may have on the relationship between two other variables.

The second volume, *Weighing Chances*, introduces probability models. Frank W. Carlborg's "What is the sample size?" treats a common and very important problem in research: evaluating the study design to choose the most appropriate observational unit for analysis. In attempting to identify "The last revolutionary soldier," A. Ross Eckler illustrates that logical examination of data, detection of outliers, and correction of errors in a data set are necessary before analysis can begin. The reader sees how frequently errors may be present even in published data sources.

In "Ratings of typewriters" in the second volume, Mosteller presents and criticizes several alternative methods of assessing and computing unusualness (deliberately avoiding introduction of the analysis of variance), and emphasizes that statistical analysis is an art as well as a science. I feel that a big problem in the public image of statistics is that people often think of statistics as a straightforward technical skill—they do not see its challenge and art.

Volume three focuses on *Detecting Patterns.* "Predicting the outcome of the World Series" by Richard G. Brown introduces the idea of solving a complex computational problem by using simulation—either by hand or with the help of a computer program. Ralph D'Agostino's "How much does a 40-pound box of bananas weigh?" illustrates the challenges involved in quality control.

Volume four is *Finding Models.* Stephen Fienberg in "Randomization for the selective service draft lotteries" uses scatterplots, 2×2 tables, chi-squared analysis, and least-squares regression to evaluate the fairness of the 1970 draft lottery.

In short, to quote the SBE preface: SBE contains "a body of material very different from the usual. ... These volumes are not mere collections of problems and examples. Rather, each represents a series of mini-learning experiences ...." Many thanks to Fred Mosteller for having led yet another worthwhile collaborative project of his fellow statisticians. In this, as in others, not only does he ensure that "everyone around him makes many important contributions to statistics" but also that the final product is a timely, well-written, and very readable document.

Weather and Climate Modification

Panel on Weather and Climate Modification, Committee on Atmospheric Sciences, National Research Council. *Weather & Climate Modification: Problems and Progress.* Washington, D.C.: National Academy of Sciences, 1973.

by
Michael Sutherland

As a member of the Panel on Weather and Climate Modification of the National Academy of Sciences, Fred Mosteller participated, along with numerous scientists, in the project that led to this report. In turn, Fred characteristically involved some of his students in the work. I was fortunate to be one of them. A master mentor, Fred turned to his students on this and many other occasions to provide raw horsepower for a multitude of projects. The experience served as a pragmatic training ground for the "apprentices" that he guided through Ph.D. programs at Harvard.

In brief, the report was one in a sequence of attempts to summarize what scientists had learned from weather modification projects around the globe. In a sense, it used a form of meta-analysis in evaluating those projects (mostly after 1950) that had a real experimental design component—treatments being randomized over "equivalent" days. Over the years I've found that weather modification is a marvelous arena in which to open up students to quantitative thinking, writing, and decision making. Three more recent and accessible references are the invited paper (with extensive comments) "Field Experimentation in Weather Modification" by Braham (1979), the text *Weather Modification by Cloud Seeding* by Dennis (1980), and a delightful and accessible *Science* article and ensuing set of letters discussing the possible costs and benefits of seeding hurricanes (Howard et al. 1972), reprinted in *Statistics and Public Policy* by Fairley and Mosteller (B41, pp. 257–294).

This informative and instructive report shows that clear organization need not suffer when a varied group of methodologists and scientists comes together to consider the state of knowledge in a complex, poorly understood, and barely explored field. Weather modification clearly had the potential for massive societal effects, ranging from food production in a hungry world to legal issues for well-to-do ranchers who feel "their" rain has been stolen. Dennis (1980) discusses what is known about the physics of cloud seeding, the evidence for its effects, and the legal issues surrounding it.

This problem area typifies Fred's involvement with the application of statistical thinking to large, complex issues that have great potential impact on man and society. The immensity of the subject area, its lack of a well-accepted theoretical framework, and the lack of textbook experimental evidence or even textbook distributions to work with gave my co-workers on the project and me a profound counterpoint to our coursework in theoretical statistics and measure-theoretic probability.

The report consists of 12 chapters organized into three broad sections:

I: Summary and Recommendations

1. Summary of Recent Work in Weather and Climate Modification
2. Recommendations for National Policies and Progress in Weather and Climate Modification

II: Technical and Scientific Advances in Weather Modification

3. Planned Modification
4. Summaries of Selected Precipitation Modification Programs
5. Review of the Modification of Certain Weather Hazards
6. A Critique of the Techniques Used in the Design, Operation, and Analysis of Weather-Modification Experiments
7. Inadvertent Modification of Weather and Climate
8. References for Part II

III. Role of Statistics in Weather Modification

9. Summary of Conference on Statistics and Weather-Modification Experiments
10. Special Topics in Statistical Analysis
11. References for Part III
12. Appendix: Summary Tables of Randomized Precipitation Modification Experiments

The publication itself has certainly stood the test of time. Rereading it brought back the immediacy and excitement of being involved in writing parts of the statistical sections. It reminded me of how much I enjoyed working on such large projects with Fred and others at Harvard during the late 1960s and early 1970s. The projects had a sense of importance and social reality that many of my friends in other fields seemed not to find in their graduate education. My work on them allowed me to integrate my personal beliefs and feelings about my society and my potential role in it with the intense, almost selfish, pleasure of doing mathematics.

Statistics has been so enjoyable for me because it involves technical skills and opportunities to bring in one's hunches, non-technical goals, and beliefs when solving problems. With Fred my graduate education combined the personal pleasure of theory with the practical challenge of work on real, fuzzy problems. Rereading this report made me realize again how much the way we "do science" and even the science that we do is intimately tied to our larger society, its needs, its perceived needs, and its politics. From my apprenticeship with Fred I learned the importance of training oneself to see beyond the immediate technical issues and pleasures of research to a larger societal context where questions of law, commerce, and moral beliefs demand equal weight. Over the years, I've carried that style to my own teaching (often using weather modification issues) and have been fortunate in finding colleagues in a variety of fields who also see the role of "professing" as more than just technical training (see Bernstein (1980) and Raskin and Bernstein (1987)).

The panel recommended three major goals and associated time frames:

1. Completion of research to put precipitation research on a sound basis by 1980;

2. Development during the next decade of the technology required to move toward mitigation of severe storms;

3. Establishment of a program that will permit determination by 1980 of the extent of inadvertent modification of local weather and global climate as a result of human activities.

From my experience with a variety of people involved in weather studies I'd have to agree with Fred's own assessment of the report's direct impact (P114): "As far as I know, nothing has come of this recommendation." Although Fred was referring to only the panel's recommendation to fund the creation of a Weather Modification Statistical Research Group, my own informal sampling has convinced me that in many ways the report has been overshadowed by the almost simultaneous SMIC Report on Inadvertent Climate Modification. Most researchers in the area of climate think I'm referring to the SMIC Report when I first mention the Weather and Climate Modification Report of 1973.

The report appeared at a time when government's willingness to fund long-term, large-scale experiments was decreasing. This was especially true in areas that had only soft estimates of a payoff. Furthermore, as a people, we were showing less willingness to be subjects in such experimentation, and various groups were demonstrating a new-found ability to legally stymie such programs. But perhaps most importantly, inadvertent human modification of the climate was being recognized as the more important societal and political issue. Issues such as whether we were depleting the earth's ozone layer or perhaps building a worldwide greenhouse or directly damaging ourselves and other living things via the by-products of our civilization caught the attention of the populace at large, the scientific community, and, ultimately, the disburser of our tax dollars—the U.S. Government.

Since the panel's report, most of our nation's research in climatology has focused almost solely on the subject of the panel's third recommendation: " ...determination ...of the extent of inadvertent modification of local weather and global climate ...." Yet, understanding inadvertent modification of weather has demanded that we first understand the weather—historically and theoretically. In a sense, the focus on inadvertent modification of the weather has led to the theoretical models, historical data, and computational simulation capabilities that lie at the core of the panel's first major recommendation: " ...to put precipitation research on a sound basis ...." Although work on actual storm mitigation has not come as far as the panel hoped in its second recommendation, I do use Fred's model of science in my own thinking and teaching about the severest storm we may ever see—nuclear winter (Harwell, 1984). Inadvertent modification or not, the elegantly modelled implications of a nuclear exchange force me to step outside the purely technical aspects and apply the only technology I have available—my vote and my voice—to try " ...to mitigate ..." such a severe storm. I'm sure I've modified the model of doing science, but it helps to start with such a firm humanistic foundation.

References

Bernstein, H. (1980). Idols of modern science and the reconstruction of knowledge. *Journal of Social Reconstruction*, **1**, 27–56.

Braham, R.R., Jr. (1979). Field experimentation in weather modification. *Journal of the American Statistical Association*, **74**, 57–104 (with discussion).

Dennis, A. (1980). *Weather Modification by Cloud Seeding.* New York: Academic Press.

Harwell, M.A. (1984). *Nuclear Winter: The Human and Environmental Consequences of Nuclear War.* New York: Springer-Verlag.

Howard, R.A., Matheson, J.E., and North, D.W. (1972). The decision to seed hurricanes. *Science*, **176**, 1191–1202. Letters and replies: **179**, 744–747 and **181**, 1072–1073.

Raskin, M.G. and Bernstein, H.J. (1987). *New Ways of Knowing: The Sciences, Society, and Reconstructive Knowledge.* Totowa, NJ: Rowman and Littlefield.

Study of Man's Impact on Climate. (1971). *Inadvertent Climate Modification.* Cambridge, MA: MIT Press.

Costs, Risks, and Benefits of Surgery

John P. Bunker, Benjamin A. Barnes, and Frederick Mosteller, editors. *Costs, Risks, and Benefits of Surgery.* New York: Oxford University Press, 1977.

by
Howard H. Hiatt

On the day in the fall of 1985 when I agreed to contribute to this volume honoring Fred Mosteller, I sat in on a faculty meeting of the Department of Health Policy and Management at the Harvard School of Public Health. As I listened to the discussion, it occurred to me that I had accepted an assignment that would not permit me to pay adequate tribute to Fred. I will comment briefly on *Costs, Risks, and Benefits of Surgery,* as I was asked to do, and with great pleasure, for it is a remarkable book and its impact has been far-reaching. However, I want also to write a few words about Fred's contributions to the Harvard School of Public Health, and particularly about his influence on its faculty.

First, the book. It is difficult to do justice to its importance, and not because the editors state at the outset, "We confidently anticipate that ... future research will produce new and better analyses, including some that may overturn some of the findings in the book, and we look forward to such work as evidence of the book's success rather than its failure." ("Heads we win, tails we win," Fred might put it.) In truth, what Fred and his co-editors might have said was that many articles in the book challenged conventional wisdom—on the "proper" treatment of breast cancer, on geographic differences in the frequency with which several operations are carried out, on surgical innovation, and on a range of other topics. The articles were intended to stimulate people to think, to ask more questions, and to seek additional answers in ways that could be useful.

The emphasis on cost containment that now surrounds many considerations of health and health policy was anticipated by the editors. They recognized long before most others that resources for health are being outstripped by medical capabilities and medical needs. The book offered some methods for analyzing the contributions of health programs and for comparing the contributions of different programs. It foresaw that the application of such methods would help facilitate rational choices.

Only a few "magic bullets" can be found in medical history; pencillin and polio vaccine are examples. Rather, the book points out, progress in diagnosis, treatment, and rehabilitation generally comes in stepwise fashion. Therefore, the grounds for enlisting doctors and patients in clinical trials are not only practical and intellectual, but ethical as well. Since all doctors and all patients are the beneficiaries of past trials, the book reminds us that all have a responsibility to participate in future ones.

Surgery is a field of far greater complexity than its technical aspects, however complicated *they* are. Therefore, in order to put together the groups for the several research projects that form the basis for the book, Professor Mosteller and his co-editors had to attract colleagues from a range of disciplines in addition to surgery and the statistical and decision sciences. Physicians from several medical specialties, economists, behavioral scientists, management experts, and others joined in. Under his creative and genial leadership experts from these disparate disciplines forged a common statement and concluded with four recommendations (pp. 392–394):

1. Appropriate studies of the effectiveness of surgical treatment should be carried out for selected conditions, particularly those where uncertainty leads to professional disagreement.

2. Our grasp of the components of cost-benefit analysis and their interrelations, the values of the various data gathering techniques, and our understanding of ethics of data gathering must be improved by theoretical and empirical work and by continued discussions in the public forums.

3. These principles of cost-benefit evaluation should be included as an integral part of the medical school curriculum; and their application to the assessment of the efficacy of medical care should be incorporated into clinical practice and continuing medical education.

4. Information on outcomes as well as costs of medical care should be routinely formulated in a manner suitable for presentation to the public.

When Professor Mosteller and his colleagues put these recommendations forward in 1977, they were regarded in some quarters as radical. But the book is not at all radical. Although it makes a highly persuasive case for what was then a profound departure from conventional practice, it does so in a moderate and responsible fashion. As Fred Mosteller has done so often in his extraordinary career, he and his colleagues in this book identify some crucial problems, describe how they have come to pass, add insights concerning their complexity and possible solutions, and end with a series of constructive proposals. Surely, the book is in part responsible for the fact that today the proposals are accepted in large measure and widely followed.

Now, back to that faculty meeting, where I happened to sit next to Fred. Before the formal discussion began, he and I talked about several important matters; the statistical justifications for the Red Sox batting order, whether neckties reflect the personality of their wearer, and aspects of modern art were among them. The business of the meeting was a review of certain courses offered by members of the Department. One faculty member discussed in detail his course in decision analysis. He described its strengths and weaknesses. He particularly spoke of how it could be more effectively integrated into the teaching programs offered by other members of the Department and by other departments. That led to a wide-ranging discussion concerning the overall teaching program of the Department; the contributions of the course in decision analysis to the program; the reactions of students; how this course would help diverse student career goals; and what it might do for the research objectives of graduate students.

These were only a few of the topics raised. Fred spoke little, but his influence was pervasive. The esprit of the department faculty, their concern for students, their interest in the work of their colleagues, their considerate questions about the possible effects of their activities on other departments, their commitment—all of these attributes were apparent. All are attributes of the Department's chairman.

Fred had come to the School in 1977 from his role as Chairman of the Department of Statistics at Harvard College. He had agreed to stay for a period of three years and then to return to Statistics. By 1980 he had more than fulfilled his understanding with the Dean. He had built a Department of Biostatistics of enormous strength. He had attracted able young and senior people from around the country. He had broadened its activities to include not only biostatistics, but the decision sciences, computer science, modeling, and other areas as well.

At the end of Fred's agreed-upon term, the Dean asked him to consider taking on the chairmanship of the Department of Health Policy and Management, a relatively new department in the School. He accepted this additional challenge. (How rarely Fred declines a request for help!) In a remarkably short period, the Mosteller leadership magic was evident in the new department, too: in a group of young faculty who believe in themselves and each other, who are deeply committed to principles of scholarship, and who work creatively in individual, departmental, and school-wide teaching and research programs. Quickness of intellect, generosity toward students and each other, good humor—these qualities are much in evidence. They are qualities present in abundance in this remarkable man. As is true for his colleagues in the Department of Biostatistics, those elsewhere in the School, and others around the world, the Health Policy and Management faculty members regard Fred Mosteller as an extraordinary model. His influence is widely felt—at the Harvard School of Public Health, elsewhere at Harvard, throughout his field, and beyond. And what a splendid influence that has been and continues to be!

Statistics and Public Policy

William B. Fairley and Frederick Mosteller, editors. *Statistics and Public Policy*. Reading, MA: Addison-Wesley, 1977.

by
Michael A. Stoto

Statistics and Public Policy is a gem of a book. Edited collections rarely tell so much about their editors. The broad range of topics exemplifies Fred's interests and work and implicitly defines the toolbox for statisticians who wish to contribute to public policy. The treatment of these topics, with its concern for lucidity, accuracy, and relevance, sets a high standard for others to follow.

The collection grew out of Fred's teaching at Harvard's Kennedy School of Government. In the late nineteen sixties, Fred, Howard Raiffa, Thomas Schelling, Richard Neustadt, Francis Bator, Graham Allison, and Richard Zeckhauser took on the responsibility of developing a new professional curriculum in public policy and management at Harvard. The result of their work was the Masters in Public Policy program, the first of its kind in the country, and the model for many newer programs. This curriculum, a public-sector version of Harvard's professional programs in law and business, emphasized the analytical techniques that were at that time becoming widespread in government work.

Between 1970 and 1975, Fred (later joined by Will Fairley) taught a statistics course in the Public Policy program. Fred and Will had some difficulty in locating examples that both illustrated statistical topics and had important connections to public policy issues. The book is based on the articles they collected during those years. Because of his interest in pedagogy, as well as his conviction that science helps policy makers, Fred has made a sustained effort to collect good examples of how statistics contributes to policy decisions. These examples, which teach as well as persuade, also play important roles in *Statistics: A Guide to the Unknown* and in *Data for Decisions.*

Statistics and Public Policy consists of individually authored papers grouped under six headings. The book is intended to supplement a standard statistics text. Using the book in the same course that Fred and Will developed at the Kennedy School, I find that the material makes statistics come alive for public policy students. The examples not only convince them that this material is useful, but also serve as vivid reminders of thorny issues.

The introduction, an essay by William Kruskal from *Federal Statistics: the Report of the President's Commission*, sets the stage by examining the contributions of statistics to public policy. In agreement with Fred's own views, the essay is upbeat and positive, with many examples of successful contributions of statistics, but at the same time constructively critical.

The second part takes up Exploratory Data Analysis, a subject in which Fred is a pioneer. Three original articles by Kennedy School colleagues Will Fairley, Bob Klitgaard, and Don Shepard exemplify the potential of this new approach to policy analysis.

Part III assembles three articles illustrating the most traditional techniques of statistical analysis in policy work. The first article, by Peter Bickel, Eugene Hammel, and William O'Connell, uses contingency tables to explore sex bias in admission to the graduate programs of the University of California—Berkeley. The article and subsequent correspondence between Kruskal and Bickel (reprinted in the book) are classics. Together they remind us that the policy answer depends on what you control for, and what you control for is a policy question that involves more than purely statistical judgment. The other articles in the section reiterate this message and at the same time illustrate age standardization, covariance adjustment, and multiple regression analysis, all widely used in policy analysis.

The fourth section begins with a unique essay (which appears only in this book) by Fred, "Assessing Unknown Numbers: Order of Magnitude Estimation." In addition to providing useful advice about how to look up a number in a reference source, collect the number in a systematic survey, or get experts to provide the number, this essay has a marvelous section on how to make informed guesses. Where else can one find a careful discussion about using rules of thumb, factoring, decomposing, triangulating, and mathematical modelling? Drawing on Fred's keen understanding of the problems faced by policy makers who use statistics, the article illustrates the value of experience and systematic analysis in topics beyond the normal range of mathematical statisticians.

The section closes with two articles about the role of experiments in evaluating social innovations. One, by John Gilbert, Richard Light, and Fred, analyzes a group of actual evaluations of social programs. This paper assesses the contributions of evaluations to policy decisions and offers some insights about how to improve the usefulness of social evaluations. Two themes emerge and illustrate Fred's interests and contribution to policy analysis: the importance of randomized experiments for drawing reliable policy conclusions and the ethics of experimentation. In this article Fred and his colleagues make the important ethical point that the alternative to controlled experimentation, in which people are haphazardly assigned to treatments with unknown risks and uncertain benefits, not only fritters away valuable experience, but is ethically repugnant because it amounts to "fooling around with people."

Part V of *Statistics and Public Policy* deals with methods for decision making. The opening paper, "The Decision to Seed Hurricanes" by Ronald Howard, James Matheson, and Warner North, comes from *Science* magazine along with the letters, comments, and rebuttals that frequently accompany the most important papers in that journal. The collection illustrates not only the careful use and persuasiveness of an important analytical technique, but also the nature of the controversy that often surrounds policy analyses. A now-classic article by Tversky and Kahneman, "Judgment under Uncertainty: Heuristics and Biases," completes the collection.

The last part of the book deals with statistical models in legal settings. The centerpiece is the California Supreme Court's decision in the famous case of People v. Collins. This case involves a "mathematical argument" about the likelihood of finding more than one couple in Los Angeles with the following characteristics: a Negro man with a beard and a white woman with a pony-tail together in a yellow convertible. This case and an accompanying note by Will and Fred take the reader through a provocative, in-depth exploration of the potential contributions of probabilistic reasoning in the courts. Additional articles by Michael Finkelstein and Will Fairley and by Laurence Tribe round out the discussion of statistical evidence.

Statistics and Public Policy describes the range of statistical techniques and ideas used in public policy analysis. The collection makes the point that simple techniques can and do make a big contribution to policy issues. From it, statisticians trained in classical mathematical statistics can learn a lot about how statistics is actually used. Policy analysis turns out to depend on different statistical ideas than we usually teach in "regular" statistics courses.

Despite the breadth of the subject matter, a common theme unifies *Statistics and Public Policy.* Many of the articles seem to have a detective-work quality to them: each step in the analysis changes the apparent conclusion and leads to deeper analyses. In the Berkeley admissions case, for instance, whenever the analysts control for a new variable, the conclusion about the apparent nature of sex bias changes. In the discussion of hurricane seeding, the original analysis seems persuasive, but the letters to the editor raise points that seem fatal until the authors dispel the doubts in their rebuttal. In People v. Collins, the use of probabilistic reasoning appears at times sensible or silly, helpful or confusing, proper or misguided. Pedagogically, these turnarounds force students to think carefully and to understand all the ramifications of the issue. In the end students realize that these matters cannot be resolved on purely statistical grounds, and that policy considerations shape both the formulation of the problem and its resolution.

Fred takes pride in a statistician's ability to think analytically and to reach concrete verifiable results. He firmly believes that statisticians and scientists can contribute to society by informing policy decisions. On the other hand, he rejects the common notion that statisticians can or should ignore their own values and the policy implications of their work. He urges statisticians to learn more about how their results are used and how to make them more useful, and to contribute more fully in the policy process. *Statistics and Public Policy* reflects Fred's views through and through. By showing us how statistical ideas and results actually contribute to the policy process, Fred and Will have helped a generation of statisticians and policy analysts to serve society better.

Data Analysis and Regression

Frederick Mosteller and John W. Tukey. *Data Analysis and Regression: A Second Course in Statistics.* Reading, MA: Addison-Wesley, 1977.

by
Sanford Weisberg

Data Analysis and Regression, by Fred Mosteller and John Tukey, holds a special place among statistics books. This book is full of the wonder of trying to learn from data, and it shows by example just what can, and what cannot, be gained from a statistical analysis. Published in 1977, the book made many of the ideas that the authors had developed over the preceding years accessible to a wider audience. It is a work worthy of study.

Several themes recur throughout this book. First and foremost the authors give advice on how to think through a statistical problem and on how to interpret results of an analysis. Few statisticians have more experience in helping policymakers and scientists use statistics than Mosteller and Tukey, and the reader can learn much from their guidance. The second theme is the combination of what we now call exploratory and confirmatory analysis, the former being more informal and graphical, while the latter uses the more traditional techniques of analysis based on sufficient statistics and tests or other inference statements. The third theme is in presenting statistical methods, both old and new, in a unified framework. These themes encompass much of what an applied statistician thinks of as important in statistics, certainly too much to catalog in a brief review. Rather, I discuss here a few thoughts on some of the contents of the book.

On Pedagogy

As a classroom textbook, this book tells us something about what Mosteller and Tukey think is important to convey to students. Repeatedly, one notices the obvious pleasure that they take in making data help answer questions of interest. The feelings of wonder and curiosity come across in several ways. First, they use real data sets, presented and discussed in a way that makes the importance and context of the data apparent. The examples are carried far enough that the reader not only learns some statistics, but also learns about the problem that the data help illuminate. The range of examples matches the wide interests of the authors, including health and medicine, public policy, social psychology, and the natural sciences. The presentation makes the examples come alive for the reader.

Throughout, the major focus is not the numbers themselves but the ideas that they represent. Fred Mosteller has used a startling demonstration in his elementary statistics course at Harvard. Early in the year, when talking about randomness, he introduces the ideas of random numbers and random number tables. He takes the Rand random number tables, *tears out a few pages*, and hands them out to the class. The effect of seeing a Harvard professor tear pages out of a book in front of a class is unforgettable (during my last year of graduate school, Mosteller's Rand tables were so depleted he had to buy a new copy!). The lesson concerning the meaning of a few numbers, as well as a better understanding of randomness, is clearly learned.

Much of the book discusses computations that can be done by hand, perhaps on the back of an envelope during a faculty meeting or a Congressional hearing. In a world where some wear calculators on their wrists and others have computers on their desks, the art of making quick calculations (like quick logs or square roots) may seem unnecessary or dull, but some among us find these to be interesting pastimes, worth learning. Clearly the regularity of numbers has always interested both authors, Mosteller's work on prime numbers (P91) and the Goldbach conjecture (P90) being two cases in point.

Cross-validation

The 1968 edition of the *Handbook of Social Psychology* contained an early version of several of the chapters in the current book (a set of page proofs for those early chapters is one of my prized possessions). One of the important ideas that Mosteller and Tukey presented in the *Handbook* article is cross-validation. The idea is simple: to validate a model fitted to data, one should divide the data into parts, and use a different part for validation than is used for estimation. A fitted model, including estimates, choice of functional form, and the like, is usually optimized for the data at hand, so the fitted model will match the data used for estimation more closely than would the "true" model. By reserving a portion of the data for model validation, the analyst can get a better idea of how well the model describes the underlying process that generated the data and in particular how well one can expect a model to predict future observations. Although this idea surely predates 1968, the presentation in the *Handbook* seems to mark the beginning of rapid development of cross-validation that we have seen since then.

Leave-one-out Methods

One of the important ideas in applied statistics, made feasible by the availability of computers, is that estimates and procedures can be repeated on subsets of the data. In the most extreme case a procedure is applied to each subset of $n-1$ cases in an n-case data set, giving a set of n estimates to study. When these n estimates are combined to give new estimates or to assess variation, we have the jackknife method. When the model fitted to $n - 1$ cases is used to predict the remaining case, a function of the prediction errors can serve as a measure of the fit of a model. This idea provides an attractive alternative to the type of cross-validation that Mosteller and Tukey suggested in 1968, and it seems to have many of the same properties.

The idea of leaving out observations and then studying changes in the analysis has spawned the "influential observation industry." Although the specifics of these methods were not published until 1977 or later, many of the key ideas appear either in this book by Mosteller and Tukey or else in their earlier writing. Several times, Mosteller and Tukey warn the reader that well-separated points can drive an analysis, and they suggest refitting without the suspect points. The algebra of adding and deleting points from a regression equation was given around 1950 by several authors, including Max Woodbury, who was then working with John Tukey (the original ideas seem to date to work by Gauss, which Hale F. Trotter translated into English in the mid 1950s, apparently at Tukey's suggestion). The "hat" matrix (projection on the column space of the predictors in a regression problem), though not actually present in Mosteller and Tukey, is almost included. Although the technology of doing linear regression has changed greatly since 1977 (my own regression textbook has been written and extensively revised in that time), most of what Mosteller and Tukey said then can be applied now.

Leave-one-out methods seem continually to be spawning new relatives. The most prominent of these is the bootstrap, which I have heard described as the West Coast jackknife. Different sorts of generalizations arise when we think of leaving out only a (small) fraction of a case or equivalently just slightly perturbing a case or a carrier. This sort of idea is familiar to numerical analysts and to differential geometers, but is new in statistics, and can be traced to the leave-one-out methods.

Statistics books often lack personality. One often finds catalogs of methods, sometimes just prescriptions for analysis applied mechanically in contrived examples. Many books tell the reader little about what the author thinks and believes, or what importance the author gives to various ideas. *Data Analysis and Regression* presents a clear alternative to these faceless textbooks. Mosteller and Tukey have presented a personal document that expresses a specific point of view, with a definite approach to the problems of making sense of data and of using linear regression. Since many of the ideas in the book are due to the authors, statisticians can be grateful that they can easily determine what Mosteller and Tukey think (or at least what they thought in 1977) about some statistical issues of importance.

Data for Decisions

David C. Hoaglin, Richard J. Light, Bucknam McPeek, Frederick Mosteller, and Michael A. Stoto. *Data for Decisions: Information Strategies for Policymakers.* Cambridge, MA: Abt Books, 1982.

by
William B. Fairley

Data for Decisions is one of the fruits of Fred Mosteller's interest in helping policymakers make better decisions, and it reflects his collaborative work with colleagues in the Public Policy Program at Harvard's Kennedy School of Government.

Uncommon good sense about information sources is what policymakers will find in the volume. The book organizes discussion of statistical data collection methods—including experiments, observational studies, sample surveys, longitudinal and panel studies, case studies, management records, official statistics, simulation, forecasting, mathematical modeling, introspection and advice, and checks on order of magnitude—around three major uses of data by policymakers: understanding cause and effect, measuring the present state of the world, and predicting the future.

Most of the chapters on methods of gathering information follow a standard format: define the method, using real examples drawn from policy issues; explain the strong and weak points of the method in comparison to alternatives; and discuss the general role it can play in the policymaker's tool kit for getting relevant, quality information.

The book belongs on the shelf of every policymaker who commissions the gathering of information or uses the results of such activities. The format and the near-independence of the chapters from each other make the book an attractive source for practitioners. A useful feature is the "Quality Checklist" for each method, which summarizes in simple questions the key criteria that good applications of the method should satisfy. Beyond picking a method suitable for the purpose at hand, the user can get a good idea of what to look for in best practice with that method. A discriminating lay reader could in fact build a fairly redoubtable position as a sponsor or consumer of data and data collection studies by putting the sources through these checklists.

Fred and his co-authors discuss some 85 actual policy examples of information gathering, conveniently listed in a second table of contents. The number shows the richness of real-life detail and the concreteness that help make the book truly accessible to the lay reader and user. Occasionally (as on page 29, where the reader is assumed to be familiar with a "regression line") more than general lay knowledge is required, but these are exceptions. Generally the text reflects Fred's longstanding interest in communicating "big ideas" of statistics to the broader audience of non-statisticians—in the present instance an influential one that we all hope will use information when they can and use it well.

The focus of the book on information gathering methods stops short of considering the paradigms of decision theory for assessing the value of information in a decision problem and of Bayesian methods for incorporating formal with personal knowledge in characterizing one's information about a quantity. I would have liked to see some comment, if only in footnotes or a short chapter, on how the book's topics for "information strategies" relate to these theoretical paradigms. But perhaps this would represent a theoretical excursion beyond the aims of the book.

An important theme in all of Fred's applied work—instantly appreciated by anyone who has had the good fortune to collaborate closely with him—is his essential pragmatism in bringing statistical tools to bear in a world of surprises and theory-defying nuances. This pragmatism shows up repeatedly in the present volume.

One manifestation is the understanding that every study will have problems and even likely errors: "We begin our treatment of errors with the more modest everyday difficulties and work up to deliberate frauds. No matter how careful a study, mistakes are likely to occur ..." (page 274 in Chapter 15, "Have Things Gone Awry?").

In the real world one must always choose between different kinds of imperfection, not between a perfect method and bad ones or between a true, precise value and an incorrect one. Thus, along with cautions about uncritical causal inferences, we get characteristically pragmatic advice about use of observational studies (page 74):

> Comparative observational studies may offer the only means of ever collecting any data at all about a treatment. Some events, such as earthquakes, are impossible to deliver as designed treatments, and certain treatments are unethical, as in many medical and social investigations. In these extremes, investigators must do their best with data from observational studies.

But Fred is uncommonly resourceful in the face of difficulties. He will not shrink from developing or encouraging his colleagues to develop new statistical methods if these are needed to meet a problem. The material in Chapter 15 on techniques for assessing the order of magnitude of hard-to-get-at numbers or for cross-checking important numbers in a study illustrates his ability to develop fresh attacks on a practical problem and his originality in stepping outside the bounds of the traditional statistical approaches to attack the problem from a very different perspective.

The book's substantial discussion (especially in Chapter 2) of the limitations of regression models for determining causal relationships among variables reflects the empirically-rooted scientific attitude that Fred brings to bear on factual investigation.

"What can you really learn about the world from a theoretical model?" he asks. A great strength of this book for the proper orientation of the lay user is its repeated emphasis on what kinds of information one can and cannot get from a variety of sophisticated, often theoretically-based statistical methods.

In courses in the Public Policy Program and at Harvard Law School on which Fred and I collaborated, we used a fanciful "General Knowledge" multiple-choice test to illustrate binomial probabilities. The questions were deliberately so obscure ("Who was the Secretary of the Modern Woodmen of America in 1927: (a) George Fox; (b) Peter Abraham; (c) John Tuttle; (d) Hamilton Fisher?") that correct answers would—one would suppose—occur only by chance, and the number correct out of 11 questions with 4 answers each would be binomial with $p = 1/4$ and $n = 11$. The amusing feature for classroom purposes was the expressions on the faces of students when they read the "surprise quiz" questions. For class, Fred prepared a review of the data resulting from the 19 test results and from tests on non-binomiality for the data that we had discussed in class. His discussion, reproduced with permission below, illustrates well the multiple dimensions of pragmatism that he brings to data and to the application of statistical methods to data:

> Well, the General Knowledge Test has passed all the statistical tests for non-binomiality that we have given it. Does that mean we think it really is exactly binomial with $p = 1/4$? No, our common sense must not entirely desert us. What we should think is that
>
> a. the actual departures are modest enough that a sample of 19 observations was not adequate to confirm them.
>
> b. for the purposes of analyzing small samples, or making inferences about such test scores, the binomial model appears to be a good approximation.
>
> c. some people really knew the answers to some of the questions. That ought to raise p above $1/4$. Why didn't it?
>
> That remains to be examined. One possibility suggested by the discussion initiated by Messrs. Geleta and Oree is that some participants did not answer every question on the questionnaire (contrary to instructions). They would then have lost their 1/4 chance of success on these omitted items. This is not quite the same problem as in the Coleman Report where I conjectured that blank papers may not always have been turned in to the central office. If something like this didn't happen, then we ought also to be concerned about the scoring. The Educational Testing Service finds that people doing hand scoring make a good many errors, and of course the atmosphere surrounding the administration of the General Knowledge Test is not one

> conducive to careful work. Other work suggests that scoring and bookkeeping errors are ordinarily made in favor of the person doing the scoring, but if this happened, it didn't show up in the grand total of right answers. Nevertheless, the administration of the General Knowledge Test has by its very weaknesses helped us to illustrate the binomial and its likely imperfections as a model.

In *Inference and Disputed Authorship: The Federalist*, Fred and David Wallace introduced the notion of the "outrageous event" that, although rare, as a practical matter could and would cause an event of interest with a certain frequency and therefore would belie any calculation of a probability of that event that was less than the probability of the outrageous event. An example which I have used is the probability of a catastrophic fire caused by an accidental fire in a liquefied natural gas tanker or in a storage tank. "Cowboy" tanker operations and sabotage are examples of outrageous events that might cause such a fire, and their probabilities are higher than some estimates of "the" probability of such an event derived from a probability model of accidents, implying that the probability of the event is not "the" probability at all. The example that Fred enjoys and that has entertained many in the telling is the probability of the event that Fred floors Muhammad Ali in a boxing match. Although Fred's own presence is far from unassuming, one quails at the picture of Fred stepping into the ring against the (now former) world heavyweight champion. Yet, if Ali in his haste has failed to tie a shoelace, and trips as he rushes from his corner, falling unexpectedly and awkwardly headlong onto the mat, he is out cold, and Fred is the winner on a technicality!

The balance and objectivity that characterize Fred's approach to issues and to the evaluation of alternative methods are evident throughout this book in its careful consideration of strengths ("merits") and weaknesses of ways to gather information. The policymaker who studies this book will be repaid manyfold by its sage guidance.

Statistician's Guide to Exploratory Data Analysis

David C. Hoaglin, Frederick Mosteller, and John W. Tukey, editors. *Understanding Robust and Exploratory Data Analysis.* New York: Wiley, 1983.

David C. Hoaglin, Frederick Mosteller, and John W. Tukey, editors. *Exploring Data Tables, Trends, and Shapes.* New York: Wiley, 1985.

by
Persi Diaconis

For over 25 years, Fred Mosteller and John Tukey have been emphasizing the data-analytic aspects of statistics. The message is: look at your data in a variety of ways. Graphics and robustness are two key words.

These data analysts have suggested thousands of novel techniques: "try this method of averaging" or "here's a better way of plotting." The ideas are often so novel that more classical statisticians have responded, "Where did that come from?" or "How do you know?"

Understanding Robust and Exploratory Data Analysis (UREDA), together with the companion volume *Exploring Data Tables, Trends, and Shapes* (EDTTS), answers many such questions by drawing on the hundreds of papers, reports, and Ph.D. dissertations whose theoretical and simulation studies lie behind the most successful exploratory techniques. The books also report new work.

The First Volume

UREDA serves two main purposes. First, it provides a very clear introduction to the most widely used EDA techniques (stem-and-leaf displays, boxplots, analysis of two-way tables by medians, and transformations). It also provides an introduction to the basics of robust estimation (M- and L-estimators of location, robust scale estimators, and confidence intervals for location). Explanations minimize the amount of jargon. There are many worked examples. Throughout, the book conveys a very practical feel for the material, along the lines of "these are the most useful things we've learned in the past 25 years."

Second, UREDA connects the new techniques with more standard procedures, discussing in classical language the available theory and results such as efficiency comparisons.

Both books are suitable for classroom use (they have a variety of exercises) with advanced undergraduate or beginning graduate students. I have used UREDA successfully at Stanford in a master's-level course aimed at a general quantitative audience.

The level of scholarship is very high—several chapters give fascinating historical details (I was surprised to see Tukey's "resistant line" traced back to Wald). They also make a determined effort to bring in the relevant part of mathematical statistics. For example, the first chapter, on stem-and-leaf displays, discusses how many "stems" to use. Toward this end, it reviews the literature on selecting the box width of histograms, tries out various rules, and makes firm recommendations.

UREDA includes two novel topics. One, the resistant straight line, appeared in the limited preliminary edition (1970–1971) of Tukey's *Exploratory Data Analysis* but not in the published version (1977). Together with extensions to multiple regression (covered in EDTTS), this is a very welcome tool. I have found its study mandatory for understanding robust regression.

The second welcome novelty is the chapter on robust scale estimation. The chapter draws on the Princeton theses of David Lax and Karen Kafadar to provide the only accessible treatment of the techniques available for this difficult problem.

Some of Mosteller's Contributions

A team of authors developed UREDA in a highly collaborative effort under the guidance and direction of Hoaglin, Mosteller, and Tukey. Table 1 lists the chapters and their authors.

Well-written chapters follow a unified connected progression. Chapters often went through many drafts. One of my interactions with these three editors was for Chapter 1 of EDTTS. At about my sixth draft, I received a completely rewritten seventh draft of a 60-page manuscript. Needless to say, my reaction was a new eighth draft, and so it went. This all took place with such care and good taste that the only possible reaction was "Hey, these guys are really serious. Oh well, back to work." My essay began as a slim article (Diaconis, 1981). Anyone interested in measuring the invigorating effect of working with Mosteller and Co. should compare the two.

The writing effort was structured around weekly meetings at Harvard's statistics department. These were attended by graduate students and local and visiting faculty. Fred Mosteller orchestrates such meetings in a no-nonsense style. You're there to get something done, be it getting deeper into scientific matters or working at the exposition. There was a constant theme of "finish the draft as best you can; then we'll try to help." Another organizational lesson I've learned from Fred: when there are competing tasks, "finish the job that's nearest done."

In thinking about what makes such a project work, one clear image shines through—somebody has to put the time in. Mosteller is simply overwhelming here. He has so much to do, so many projects he's involved in. Yet somehow he manages a careful reading of your draft and usually sends an

TABLE 1. Contents of UNDERSTANDING ROBUST AND EXPLORATORY DATA ANALYSIS

1. *Stem-and-Leaf Displays*, John D. Emerson and David C. Hoaglin
2. *Letter Values: A Set of Selected Order Statistics*, David C. Hoaglin
3. *Boxplots and Batch Comparison*, John D. Emerson and Judith Strenio
4. *Transforming Data*, John D. Emerson and Michael A. Stoto
5. *Resistant Lines for* y *versus* x, John D. Emerson and David C. Hoaglin
6. *Analysis of Two-Way Tables by Medians*, John D. Emerson and David C. Hoaglin
7. *Examining Residuals*, Colin Goodall
8. *Mathematical Aspects of Transformation*, John D. Emerson
9. *Introduction to More Refined Estimators*, David C. Hoaglin, Frederick Mosteller, and John W. Tukey
10. *Comparing Location Estimators: Trimmed Means, Medians, and Trimean*, James L. Rosenberger and Miriam Gasko
11. *M-Estimators of Location: An Outline of the Theory*, Colin Goodall
12. *Robust Scale Estimators and Confidence Intervals for Location*, Boris Iglewicz

encouraging letter (or 2 or 3!). If he can put the time in, so can you.

The group works hard on being positive, and many a stalled project has been put back on course by a late-night call from Mosteller: "I just called to tell you how pleased I am with where it's going."

The Second Volume

Exploring Data Tables, Trends, and Shapes (EDTTS) is an effective second course in EDA. It brings the basic techniques to life in larger-scale problems like multiple regression and multiway contingency-table analysis. EDTTS contains more new material than UREDA. Two chapters that struck me as particularly noteworthy discuss "the square combining table" and graphical methods to assess goodness-of-fit for discrete frequency distributions.

The square combining table offers an alternative method of obtaining a resistant fit to a two-way data table. It proceeds by computing and summarizing differences between each pair of rows and between each pair of columns. These summaries are combined, using an approach developed for paired comparisons, to yield row and column effects.

The chapter on checking the shape of discrete distributions offers methods of plotting discrete data which give straight lines (with slope depending on the unknown parameter) if the model is correct. The methods are laid out for common exponential families with emphasis on Poisson and binomial examples.

Fred Mosteller is a co-author of two chapters in EDTTS. The first, joint with Anita Parunak, discusses methods of identifying outliers in large contingency tables. It contains many bright new ideas, tried out in real and simulated problems. It's vintage Mosteller—an important practical problem sensibly attacked. The paper opens a new corner of research; but if I had such a table, I'd try their suggestions and feel I'd probably squeezed out most of the juice in that direction.

Fred's second chapter is joint with Andrew Siegel, Edward Trapido, and Cleo Youtz. It presents an important contribution to what may be called the psychology of data analysis. They carried out a study to learn how we fit a line to a two-dimensional point cloud when we fit by eye. The students who participated preferred a line closer to the point cloud's major axis than to the least-squares line.

Table 2 summarizes the titles and authors of the chapters in EDTTS.

In summary, these two books have made a great contribution to statistics. They are filled with the wisdom of years of thought, experimentation, and practical trial. With unusual clarity, they simultaneously make data analysis come alive for the outsider and for the mathematical statistician.

TABLE 2. Contents of EXPLORING DATA TABLES, TRENDS, AND SHAPES

1. *Theories of Data Analysis: From Magical Thinking through Classical Statistics*, Persi Diaconis

2. *Fitting by Organized Comparisons: The Square Combining Table*, Katherine Godfrey

3. *Resistant Nonadditive Fits for Two-Way Tables*, John D. Emerson and George Y. Wong

4. *Three-Way Analyses*, Nancy Romanowicz Cook

5. *Identifying Extreme Cells in a Sizable Contingency Table: Probabilistic and Exploratory Approaches*, Frederick Mosteller and Anita Parunak

6. *Fitting Straight Lines by Eye*, Frederick Mosteller, Andrew F. Siegel, Edward Trapido, and Cleo Youtz

7. *Resistant Multiple Regression, One Variable at a Time*, John D. Emerson and David C. Hoaglin

8. *Robust Regression*, Guoying Li

9. *Checking the Shape of Discrete Distributions*, David C. Hoaglin and John W. Tukey

10. *Using Quantiles to Study Shape*, David C. Hoaglin

11. *Summarizing Shape Numerically: The g-and-h Distributions*, David C. Hoaglin

References

Diaconis, P. (1981). Magical thinking in the analysis of scientific data. *Annals of the New York Academy of Sciences*, **364**, 236–244.

Tukey, J.W. (1977). *Exploratory Data Analysis.* Reading, MA: Addison-Wesley.

Beginning Statistics with Data Analysis

Frederick Mosteller, Stephen E. Fienberg, and Robert E.K. Rourke. *Beginning Statistics with Data Analysis.* Reading, MA: Addison-Wesley, 1983.

by
John D. Emerson

On a visit to Fred Mosteller's office in the Harvard Science Center a few years ago, I noticed large loose-leaf notebooks on his bookshelves with the lettering *BEST* running down the spines. I correctly guessed that *BEST* must be an acronym for yet another of Fred's team projects. Fred Mosteller is a master at organizing, nurturing, and completing projects of research and exposition that involve from two or three people to more than a dozen. As it turned out, *BEST* referred to a new kind of statistics textbook that Fred, Steve Fienberg, and the late Bob Rourke were writing. It was to be named *BEginning STatistics with Data Analysis.*

Like numerous other projects that Fred has inspired, the *BEST* project bore fruit to be savored by many. This text represents a marked departure from those that preceded it, and many of its good features are imitated by several texts just now appearing.

I find it difficult to characterize the level of *Beginning Statistics* or to define a single audience for which it is best suited. Certainly, the mathematical level is modest: no calculus is needed, and at least the first half of the book should be accessible to students with a year of high school algebra. Later chapters introduce algebraic notation as needed; summation notation, subscripts, and hats are used freely. Even so, the exposition of multiple regression, analysis of variance, and nonparametric statistics is as gentle as possible while offering mature insights about these topics. *BEST* could serve well a variety of audiences ranging from bright high school seniors, through graduate students from diverse areas that need statistical methods, to experienced workers in industry, the health professions, or scientific laboratories.

Among the many features that combine to give this fine book its uniqueness and its considerable stature among beginning statistics texts, I would highlight the following:

- *BEST* provides a comprehensive view of statistics as a broad science; it introduces and explains many of the broad concepts whose understanding and appreciation are so essential to statistical practice.

- *BEST* virtually overflows with captivating and provocative examples that use real data.

- *BEST* finds a good balance between modern data analysis and statistics; some of the most useful ideas and methods from exploratory data analysis (EDA) are introduced in early chapters, and a data-analytic viewpoint is carefully interwoven through much of the text.

- The many illustrations are amply motivated, carefully integrated throughout the text, and made to pay off generously.

- Mathematical formalisms and probability theory appear only when needed for the statistical development to proceed; as a result, they emerge in an unobtrusive and non-intimidating way.

I now expand on and illustrate each of these features.

BEST provides an impressively comprehensive (for a beginning text) view of statistics as a science; it goes far beyond the standard treatments of statistical methods or the newly proliferating treatments of data analysis. For example, an early chapter discusses "Gathering Data" (case studies, sample surveys and censuses, observational studies, controlled field studies and experiments), and the final chapter examines "Ideas of Experimentation" (stratification, randomization, matching, blocking, covariates, blindness, loss of control, causation, initial incomparability, confounding). These discussions, along with many others, effectively convey insights and perspectives that grow out of Fred's broad experience in statistical applications across many disciplines.

An index to the data sets and examples in *BEST* shows 92 entries; many of these are discussed on more than one occasion and at increasing levels of insight as the exposition of the text unfolds. Among the illustrations that my students and I have found most intriguing are: the distribution of police calls in New York City over the hours of a day, the relationship between reported sinkings and actual sinkings of German submarines by the U.S. Navy in World War II, the Pennsylvania Daily Lottery and its possible fixing, and a potential relationship between cancer mortality and fluoridation.

BEST incorporates exploratory data analysis in a way that enhances the presentation of traditional statistical topics. The first four chapters introduce and illustrate techniques for exploring single batches of data, y-versus-x data, several batches at different levels, and two-way tables. The concepts of a model, fit, and residuals are firmly established with examples, long before inference-related concepts like distribution are encountered. Examples introduced in these chapters sometimes reappear in later chapters—for example, regression is used to examine data on predicted and actual sinkings of German submarines. I especially appreciated the early introduction of some basic structures of data, and the questions implicit in each, long before suitable methods for formal statistical analysis are developed.

Beginning textbooks in statistics sometimes leave an impression that the examples are afterthoughts and only facilitate understanding a statistical technique and its application. *BEST* conveys a very different message: the examples are presented as the *raison d'être* for the statistical methods. Just how well *did* the Navy do in reporting the number of sinkings of German submarines? Why were the sinkings somewhat underreported when one might expect the military to overstate its accomplishments in battle? Those of us who have worked with Fred know well his insistence that we always address the questions, "What is the payoff? What does the analysis tell us about what may really be going on?" My students, like me, were fascinated to learn about a lottery that used ping pong balls which may have been injected with paint using a hypodermic needle. Or to be guided through a regression analysis that ultimately shows how the age distribution of populations in various cities explains away an apparent link between water fluoridation and cancer mortality. Each attractive example substantially enhances one's interest in the development of statistical methods.

I've always enjoyed teaching probability theory, and I view it as a foundation upon which statistical inference rests. Having on several occasions drawn material from Fred's delightful *Fifty Challenging Problems in Probability*, I was surprised that the word "probability" makes its first appearance on page 162 of *BEST*. A single (and optional) chapter, the eighth, gives a gentle and intuitive treatment of probability theory. Subsequent chapters use language like, "Let p_1 be the true proportion of volunteers who would get no colds if we have infinitely many volunteers using vitamin C, and let p_2 be the corresponding true value if we have infinitely many taking placebos." My students learned to understand and use probability-based arguments without hours of instruction in combinatorics and counting and in formal probability theory. Although my own bias is for more emphasis on probability, I do know that *BEST* worked well for my students. They developed a feeling for probability and its role in statistics, and they had more time to devote to the many other important issues that comprise statistics.

Similarly, although the algebraic and notational level of *BEST* increased dramatically from the first chapter to the latter ones, the shift was gradual and quite imperceptible. Fred Mosteller has a knack for "bringing the reader along" and causing more intellectual growth to occur than one might think possible. I believe that the inclusion of "conceptual formulas" in parallel with algebraic equations contributes to effective pedagogy. Such formulas as

$$\text{variance of } \bar{p} = \left(\frac{\begin{array}{c}\text{population}\\ \text{variance}\end{array}}{\text{sample size}} \right) \times \left(\begin{array}{c}\text{improvement factor}\\ \text{for sampling}\\ \text{without replacement}\end{array} \right)$$

and

$$\text{critical ratio} = \frac{\text{observed statistic} - \text{contemplated value}}{\text{standard deviation}}$$

aid students in grasping concepts that must transcend algebraic developments.

BEST directly confronts difficult statistical concepts and issues, even as it avoids letting probability or mathematical formalism become barriers to understanding. For example, it explains the concept of degrees of freedom in the calculation of a sample variance with these sentences: "We have used the sample $\bar{x}$ as the location from which to measure deviations in computing the sample variance, and so we have arbitrarily arranged to make the sum of the deviations $(x - \bar{x})$ add up to zero, which it ordinarily would not do if we used the location value μ. This constraint on the deviations is called a loss of a degree of freedom." Similarly an optional (starred) section on predicting a new y value for regression contains an excellent discussion of the three sources of variability in the prediction.

On balance, *BEST* is the best beginning statistics textbook I have encountered. Any student who masters most of its content will have gone far toward understanding statistics as it is applied today. I also recommend *BEST* for supplemental reading in my advanced undergraduate course in mathematical statistics; its richness of examples, breadth of coverage, and depth of statistical insight are indeed remarkable. Fred Mosteller and his coauthors have made an innovative and welcome contribution to the teaching and learning of beginning statistics.

Biostatistics in Clinical Medicine

Joseph A. Ingelfinger, Frederick Mosteller, Lawrence A. Thibodeau, and James H. Ware. *Biostatistics in Clinical Medicine.* New York: Macmillan, 1983.

by
Joel C. Kleinman

Biostatistics in Clinical Medicine (BCM) is a landmark in introductory biostatistics textbooks. Although several good texts are available, BCM provides a unique combination of sound statistics, clear explanations and, most important, incredibly good motivating examples of how and why statistical thinking is essential in clinical medicine. The widespread adoption of this book as a text for medical students could whet enough appetites to help fulfill the Nation's critical need for clinical investigators and epidemiologists.

Chapter 1 sets the tone immediately. Rather than beginning with statistical concepts, it takes up one of the most common clinical problems: how should the results of diagnostic tests influence the physician's beliefs about the patient's disease? Concepts such as sensitivity, specificity, and predictive value of a diagnostic test are then introduced. Using these concepts and a specific clinical example as motivation, the authors define and illustrate probability, conditional probability, independence, and Bayes's Theorem. The chapter ends with a more realistic example involving an algorithm for multiple testing to diagnose pancreatic cancer.

Chapter 2 extends the diagnostic testing examples to introduce the concepts of prior and posterior probabilities, odds, likelihood ratio, and the binomial distribution. The authors present the simple formula

$$\text{posterior odds} = (\text{prior odds}) \times (\text{likelihood ratio})$$

by which the patient's odds of having the disease can be modified according to the diagnostic test's likelihood ratio. They give clear, intuitive definitions and explanations of each term in this formula.

Chapter 3 applies the probabilistic concepts introduced earlier to decision trees. The authors develop a complex example for the management of patients with urinary tract symptoms. Many students exposed to these methods for the first time often argue that it is impossible to assign numerical scores to reflect relative values of disparate outcomes (e.g., life with one leg, life with two legs, and death). The authors anticipate this argument with an example which clearly illustrates that any decision has implications about the relative values of such outcomes. They also encourage sensitivity testing to examine the decision alternatives when probabilities or weights assigned to the outcomes change.

Chapter 4 introduces the central idea of variability by considering the clinical problem of evaluating the effectiveness of therapy for a patient with angina. The authors introduce a useful three-level classification of sources of variation: biologic, temporal, and measurement. Chapter 5 treats the ideas of variability in more detail, using blood pressure measurement as an example. The Gaussian distribution, t-distribution, standard deviation, confidence limits, and significance tests are all introduced. Chapter 5A (optional) develops the ideas of statistical inference more formally. Chapter 5B (also optional) presents exploratory data analysis techniques in the context of analyzing variability with special emphasis on outliers.

In Chapter 6 the authors continue the example of assessing treatment efficacy but shift from a continuous outcome measure (blood pressure) to counts. Confidence intervals and significance tests are considered for the binomial and Poisson distributions. In an interesting change of pace, the authors compute 80 percent confidence limits instead of 95 percent. They cite two reasons for the change (p. 139):

> first, because in dealing with single patients, as opposed to many, data come slowly, and one may wish to ease up on the confidence required; and second, we want to emphasize that there is nothing sacred about the 95% level.

This change is important because it illustrates that conventional statistical techniques should not be automatically applied in all situations. The practical situation faced by the investigator (and the implications of Type I versus Type II errors) needs careful attention in any application of statistics. The explanations and graphical illustrations in this chapter are particularly helpful.

Chapter 7 explains P values in greater detail. This extremely valuable discussion contains a great deal of sound practical advice on the utility and limitations of P values. The authors illustrate the problems of interpreting P values when multiple comparisons are made and explain how Bonferroni's inequality can sometimes help the interpretation.

The discussion of the role of power is particularly cogent. The clinical implications of power calculations are effectively highlighted by the authors' suggestions (p. 173) that

> When using significance tests in the clinical setting, it makes sense to relax the significance level, for instance, to .10 or .20.
>
> In this way, large effects can be recognized as possibly real, allowing the possibility for further inquiry to separate real effects from false positives. The alternative strategy, routine use of the .05 criterion, essentially ignores the costs of insensitive testing.

The authors go on to discuss the importance of presenting power calculations in the report of results. Their example is instructive (p. 174):

> In an investigation of size $n = 5$, if 2 successes occurred, we might report the 95% confidence interval of .08 to .81. But this is rather different from telling the reader that the investigation as designed had only 1 chance in 6 of detecting an improvement in success rate from 20% to 30% percent. Thus the reader may find a somewhat different impact from the two statements.

This chapter should be required reading for extremists on either side of the artificial controversy now raging among epidemiologists regarding the role of confidence intervals versus P values. Clearly, a complete report of a statistical investigation should contain both (together with power statements).

Chapter 8 (optional) describes chi-square statistics. Since the chi-square distribution rarely arises in analyzing data from an individual patient, the authors shift the example to a physician's assessment of success in treating hypertensive patients.

Chapters 9 and 9A are excellent presentations of the basic concepts of regression. Sections 9–2 and 9–3 are particularly important in that they show how regression to the mean can lead to erroneous conclusions regarding efficacy of therapy.

Chapters 10 and 11 provide a most appropriate climax to this text. Most physicians will not become clinical investigators themselves. The most important reason for a medical student to learn biostatistics is to read reports of clinical investigations and make judgments about the quality of the evidence presented. Chapter 10 points out what to look for in a report of a clinical trial in order to assess whether the treatment had an effect. The authors make a strong case for randomized trials. They present an unusually compelling example of the potential bias introduced by unmeasured variables in nonrandomized trials (p. 227):

> As part of the Coronary Drug Project, over 8300 men with a history of myocardial infarction were assigned in a random manner to various therapies. Over 2700 were assigned to placebo. Those patients who actually took the placebo had about one-half of the five-year mortality of those patients assigned to placebo who did not take at least 80% of the prescribed medication. This striking difference persisted even when mortality was adjusted on the basis of 40 prognostic factors in the two groups of patients. It seems probable that many factors associated with pill-taking behavior influence mortality from coronary disease, factors we do not measure at present. (p. 227)

The authors go on to provide extremely informative discussions and practical advice concerning the roles of blinding, complete followup, consequences of imbalance in randomly generated treatment groups, adjustment (and over-adjustment), multiple comparisons, interpretation of nonsignificant P values, and the implications of inaccurate diagnoses and suboptimal therapy. They present important advice regarding the need to count all relevant events in a trial but balance the resulting loss of power against practical considerations (p. 238):

> When trials are used as guides to therapy, all clinically important events should be reported. This may reduce the statistical power of the investigation, but the investigation will provide an estimate of the total risks and benefits associated with therapy. We may have to be satisfied with results that show a trend toward total benefit and a statistically significant reduction in a specific type of morbidity or mortality.

In the final chapter, the authors consider a problem which has received little attention by biostatisticians or clinical investigators: how can the individual physician determine whether the trial's results are likely to apply to the treatment of an individual patient? There are no clear-cut approaches to answering this question, but the authors provide a great deal of insight into how the physician should think about the issue. They also provide very interesting tables concerning the probability of a chance reversal of a treatment effect in the least favorable stratum. These tables illustrate the danger of selecting a particular characteristic of the patient which puts him or her in the stratum with smallest treatment effect. For example, even with a strong treatment effect of 4 standard deviation units, there is a 50 percent probability that the effect is reversed in the least favorable of 8 strata.

Also, a helpful discussion examines the question that many students will raise when first exposed to statistical ideas: if the new treatment was better than the standard, why not adopt it even when the results are nonsignificant? The authors provide useful guidance by distinguishing the decision that the individual physician must make for the individual patient (if the treatment is safe, it's reasonable to try it out, especially if standard treatment has failed) and the policy decision that must be made in the face of a stream of treatment innovations, many of which are unfavorable and costly.

I have attempted to highlight in this discussion many of the features which make BCM an important contribution to the already crowded field of introductory biostatistics texts. Although I have no personal knowledge of the relative contributions of each author to this book, it is clear to me that its thrust and style were heavily influenced and inspired by Fred Mosteller. Even Fred's genius for small but very helpful innovations is apparent (p. viii):

> The tables have special lists on each page intended to speed the reader to the table being sought. Each page lists at the top all the tables, and shows in boldface type the table on that page.

> The reader can thus tell whether it is necessary to leaf backward or forward to find the desired table.

Although the text is directed toward the clinician, I believe it could serve as a useful introduction for nonclinical statistics courses. Even if the reader is not a physician, he or she is likely to have been a patient and will find the examples interesting and realistic. Furthermore, students of public health or health administration will find the examples especially valuable in developing an appreciation of the uncertainties that physicians face in interpreting and applying clinical investigations to patient care.

Finally, I believe that even experienced biostatisticians will benefit from reading this book. Some of the results presented (e.g., the probability of treatment reversal in the least favorable stratum) are not well known, and they provide important insight into problems of clinical research. Furthermore, the examples and explanations can greatly help consulting statisticians in improving the quality of their interaction with physicians and in providing ammunition for arguments about the need for randomized trials.

One of Fred's trademarks in the large volume of his written work is that his treatment of even the most elementary topic contains some new insight or nuance of presentation which provides a refreshing departure. *Biostatistics in Clinical Medicine* continues this tradition of innovation and excellence.

Person Index

Subject Index

MIX
Papier aus verantwortungsvollen Quellen
Paper from responsible sources
FSC® C105338
FSC
www.fsc.org

If you have any concerns about our products,
you can contact us on
ProductSafety@springernature.com

In case Publisher is established outside the EU,
the EU authorized representative is:
Springer Nature Customer Service Center GmbH
Europaplatz 3, 69115 Heidelberg, Germany

Printed by Libri Plureos GmbH
in Hamburg, Germany